Wes McGee · Monica Ponce de Leon
Editors

Robotic Fabrication in Architecture, Art and Design 2014

Foreword by Johannes Braumann and Sigrid Brell Cokcan,
Association for Robots in Architecture

with contributions by Aaron Willette

 Springer

Editors
Wes McGee
Monica Ponce de Leon
Taubman College of Architecture
    and Urban Planning
University of Michigan
Ann Arbor, MI
USA

Funded by KUKA Robotics and the Association for Robots in Architecture

ISBN 978-3-319-04662-4   ISBN 978-3-319-04663-1   (eBook)
DOI 10.1007/978-3-319-04663-1
Springer Cham Heidelberg New York Dordrecht London

Library of Congress Control Number: 2014933048

Printed on acid-free paper

Springer is part of Springer Science+Business Media (www.springer.com)

Foreword by the Association for Robots in Architecture

When the Association for Robots in Architecture was founded in 2010, just a few institutions in the world utilized robots in a "creative" context. While the works of pioneers such as Gramazio and Kohler were already widely published in architecture and design media, only a few selective clusters of creative robotic research existed, but no real network to foster collaboration and the exchange of ideas. Architects and designers considered robots to be machines that are capable of doing great things in the hands of engineers and researchers, rather than tools that can facilitate or even inform their own work in the near future. Thus, the purpose of the Association for Robots in Architecture was clear from the beginning; to "make industrial robots accessible to the creative industry". We pursue that goal with two parallel strategies: On the one hand, by developing the software KUKA|prc for easy robot control within a CAD environment, and on the other hand by acting as a network and platform toward an open access to robotic research.

Following more than a year of preparation, the first conference on robotic fabrication in architecture, art, and design—Rob|Arch 2012—took place in December 2012 in Vienna. Initially conceptualized as a symposium with a few dozen participants, it quickly turned out that there was significant interest from both academia and industry. Eight internationally renowned institutions joined us by offering two-day robot workshops—instead of just talking about the results of robotic fabrication, the robot labs were opened to the public for the very first time, giving participants an insight into the processes and workflows that usually take place in closed research labs. Also the robot industry realized the potential of new, creative robotic applications, with KUKA acting as the main conference supporter, alongside the sponsors ABB, Stäubli, Schunk, Euchner, Zeman, and splineTEX. Finally, more than 250 people attended the conference, with around 100 of them actively participating in the robot workshops.

The effects of Rob|Arch 2012 can still be felt, in the form of collaborations, business deals, and also friendships. Still, within the 18 months that have passed between Rob|Arch 2012 and Rob|Arch 2014, the robotic landscape of the creative industry has grown—and changed—rapidly. Many universities have acquired both small and larger robots, building upon existing plugins for Grasshopper to rapidly introduce their students to programming complex machines. At the same time, an increasing number of artists, architects, and designers are starting to see robotic arms as valuable design tools, while innovative firms in the classic automation business are observing the benefits of new, design-driven strategies for controlling robotic arms. This development is mirrored in the member-list of the Association for Robots in Architecture: While two thirds of the members are universities, the remaining third is made up by individual artists, fablabs, and offices, but also enterprises like Absolut and Boeing. Looking forward to Rob|Arch 2016, this ratio may approach 50/50.

Rob|Arch 2014, and this book, are representative of these changes, spanning the wide range from Google's Bot & Dolly, using robots in cinema, to highly technical robotic applications depending on sensor-based feedback in the contributions from industry partners KUKA, ABB, Stäubli, and Schunk. While in 2012 European institutions hosted universities from the United States, this year the University of Michigan and workshop co-host Carnegie Mellon University collaborate with partner-institutions from Germany, Australia, Spain, and Austria, while Princeton University is teaming up with a university spin-off, Greyshed.

Since the very beginning, the use of robotic arms has been a collaborative effort involving many "trans" disciplines. Rob|Arch 2014 again fosters the exchange of ideas not only between researchers, but also between all kinds of professionals, hackers, artists, and enthusiasts.

We want to thank the editors and conference chairs Wes McGee and Monica Ponce de Leon, as well as their entire team, for their hard work in making Rob|Arch 2014 happen. Furthermore, we want to congratulate all workshop institutions for sharing their ideas and workflows, which is most valuable for the whole community in regards to *open access* and a rapid knowledge transfer. Finally, we are grateful for the generous support of our industry partners, who do not only support the funding of the conference and the workshop infrastructure, but also devoted themselves to supporting young potentials and talents in this new field through the *KUKA Young Potential Award* and the *ABB Mobility Grant*.

We hope to see you all again at Rob|Arch 2016!

Sigrid Brell-Cokcan
Johannes Braumann

Preface

The work presented in this book exhibits the continuing evolution of robotic fabrication in architecture, art, and design. Once the domain of only a handful of institutions, the application of robotic technologies in these disciplines is consistently growing, led by interdisciplinary teams of designers, engineers, and fabricators around the world. Innovators in the creative disciplines are no longer limiting themselves to off-the-shelf technologies, but instead have become active participants in the development of novel production methods and design interfaces. Within this emerging field of creative robotics a growing number of research institutions and professional practices are leveraging robotic technologies to explore radical new approaches to design and making.

Over the last several decades there has been a widely discussed adoption of digitally driven tools by creative disciplines. With designers seeking to push the limits of what is a possible using computational design, parametric modeling techniques, and real-time process feedback, industrial robotic tools have emerged as an ideal development platform. Thanks to advances by established manufacturing industries, the accuracy, flexibility, and reliability of industrial robots has increased dramatically over the last 30 years. The accessibility of the technology to new users has also increased dramatically, with many manufacturers adopting open standards for connectivity and programming. Designers have taken the flexible nature of industrial robotic technology as more than just an enabler of computationally derived formal complexity; instead they have leveraged it as an opportunity to reconsider the entire design-to-production chain.

This is not to say that industrial robots have become mainstream. As with all digital technologies that have entered into creative disciplines, the development of knowledge surrounding the use of robotic fabrication methodologies is ongoing. And while the productive impact of their possibilities and resistances on these disciplines remains an exciting and contested territory, they have had a palpable effect that is actively shaping contemporary discourse.

Rob|Arch

Initiated by the Association for Robots in Architecture as a new conference series focusing on the use of robotic fabrication within a design-driven context, Rob|Arch—Robotic Fabrication in Architecture, Art and Design, provides an opportunity to foster a dialog between leading members of the industrial robotic industry and cutting-edge research institutions in architecture, design, and the arts. In December 2012, the first conference was hosted by its founders Sigrid Brell-Cokcan and Johannes Braumann in Vienna, Austria; now in its second iteration the 2014 conference travels to North America, hosted by the University of Michigan Taubman College of Architecture and Urban Planning. The Taubman College is well known as an academic institution for its diverse and multifaceted approach to design education, as well as its long-standing traditions in pursuing making as a form of knowledge creation.

One of the features of the Rob|Arch conference series is its focus on fabrication workshops, where leading research institutions and creative industry leaders host workshops lead by collaborative teams from around the globe. For the 2014 conference workshops there was an open call for proposals, with eight workshops selected to be held at the University of Michigan, Carnegie Mellon University, and Princeton University. Many of the workshops are based on cutting-edge work currently in progress, and their accompanying texts are published in the "Workshop Papers" section of the book.

The selected workshops cover a wide range of experimental robotic fabrication processes. The contribution from the Institute for Computational Design at University of Stuttgart focuses on their novel methodology for the production of wound composite components using cooperative robotic manipulators to produce variable units from reconfigurable tooling. A collaborative team from the University of Technology, Sydney and the University of Michigan is investigating robotic bending, cooperative assembly, and welding toward the production of complex architectural components. A workshop taught by a collaboration between the University of Michigan and IAAC focuses on sensing and material feedback within a cooperative robotics workcell. Bot & Dolly, one of the Industry Keynotes for 2014, will lead a workshop on procedural fabrication that showcases their innovative control software. Bot & Dolly is design and engineering studio that specializes in automation, robotics, and filmmaking. At Carnegie Mellon University's dFab Lab one workshop will couple cooperative robotic steam bending with integrated sensing techniques, while a team from the University of Innsbruck and the Harvard GSD will lead a workshop utilizing cooperative manipulators for the development of novel building components using phase change polymers. A third workshop at CMU will be led by a team from the Harvard GSD and TU Graz on the sensor-informed fabrication of reformable materials. And last, but not least, Princeton University will host a workshop on augmented materiality, using real-time sensor feedback and custom hardware interfaces to explore the closed-loop fabrication of structurally-optimized components.

Reflecting on the workshop and scientific paper submissions a number of themes emerged that will define both this year's conference and the near-future of robotic fabrication research, many paralleling the state of robotics and automation in other manufacturing industries. Sensor-enabled processes and robotic vision are addressed in a number of papers, both as techniques for in-process tolerance gauging and as adaptive path-planning tools. From the exploration of sensor enabled on site construction techniques, to new techniques for digitally controlled metal forming, designers and architects are expanding the capabilities of the tools at their disposal. Additionally, research projects involving cooperative robots are becoming more common, as research labs around the world have invested in multirobot work cells. This can be viewed as an indication that robotic fabrication research in architecture and design is about much more than just the subtractive or additive techniques analogous to traditional CNC processes: researchers are actively developing production methods which represent entirely new paradigms for fabrication. This is not to suggest that novel work on additive, subtractive, and material forming processes is not occurring; on the contrary, a number of papers address these topics, at scales ranging from the size of a building component, to a mobile platform capable of reaching the scale of a building.

One aspect that has been critical to this adoption has been continued focus by researchers and designers to challenge the norms of standard industrial workflows and machine interfaces. Such research continues to be a key aspect of advancing the possibilities for robotic technology to empower the design process. What is significant, however, is that robotic tools are enabling designers and architects to develop processes that suit the material, scalar, and tectonic needs of their discipline. Robotic technologies provide the ideal platform for developing fabrication processes in an experimental, iterative framework, without reinventing the machines of production.

Perhaps the most exciting trend in the field has been the growing level of knowledge transfer occurring between researchers, designers, and industry partners. The integration of robotic technologies into the workflows of creative industries has demanded renewed levels of cross-disciplinary collaboration. To further this exchange, industry partners were invited to submit papers documenting recent projects in the context of their value to art, architecture, and design. Their submissions illustrate the diversity of research and development going on in the industry, from force-control and adaptive gripper applications demonstrated by Schunk, to lightweight robotic systems by KUKA, dedicated material removal robots by Stäubli, and linked kinematic handling with cooperative robots by ABB.

As new technologies are developed across a wide range of robotic industries, innovators in the creative disciplines will continue to adapt and transform these tools to suit their specific applications. This is more than simple technology transfer, however, as robotic technologies are having a visible impact on the discourse surrounding the means and methods of production in the creative industry. Around the world this discourse is shaping not only how designers look at fabrication technologies, but the entire methodology by which they engage design and material production. As creative industries continue to explore and develop new applications for robotic technology, we look forward to new innovations enabled through collaboration between industry, academia, and the growing community of designers, programmers, and trendsetters surrounding "Robots in Architecture."

The conference chairs would like to thank the CEO of the KUKA Robot Group, Stu Shepherd and Alois Buchstab of KUKA Roboter GmbH who devoted themselves to make this conference and scientific book possible, ABB for their main support of the workshops together with Stäubli and Schunk, as well as our advisory board, and the Association for Robotics in Architecture for the opportunity to organize the conference. In addition we would like to thank the Scientific Committee, composed of architects, engineers, designers, and robotic experts; without their help it would not have been possible to develop the quality of work presented within. Special thanks to our assistant editor, Aaron Willette, for his tireless support. An especially important thanks goes to the entire team at the Taubman College of Architecture and Urban Planning, including both staff and faculty, who have supported the development of the conference. We would also like to thank our peer institutions who graciously agreed to host workshops: Carnegie Mellon University and Princeton University. Finally, special thanks to Springer Engineering for their assistance in editing and publishing these proceedings.

Wes Mcgee
Monica Ponce de Leon

Contents

Part I
Scientific Papers

Variable Carving Volume Casting

A Method for Mass-Customized Mold Making

Brandon Clifford, Nazareth Ekmekjian, Patrick Little
and Andrew Manto

Abstract The digital era fosters variability and change, though this desire loses traction when applied to methods falsely assumed to be repeatable—casting. This collision has produced a plethora of expensive, wasteful, and time-intensive methods. This chapter presents a method for rapidly carving variable molds to cast unique volumetric elements, without material waste. This method employs a multi-axis robotic arm fitted with a hot-knife to carve foam into mass-customized negatives. In doing so, it re-engages a gothic craft tradition of producing unique volumetric architectural elements. The act of rapidly carving volumetric material mines knowledge from the past in an effort to create novel forms that are not possible in the aggregation of standard building components. This chapter advocates for, prototypes, and analyses this variable, sympathetic, and reciprocal approach that carving once offered the built environment. We found the method to be effective and promising, when informed by limitations and constraints embedded in the process.

Keywords Robotic fabrication · Multi-axis · Formwork · Mass customization · Digital craft · Free-form geometry

B. Clifford (✉) · N. Ekmekjian · P. Little · A. Manto
Massachusetts Institute of Technology, Cambridge, MA, USA
e-mail: bcliffor@mit.edu

N. Ekmekjian
e-mail: nazareth@mit.edu

P. Little
e-mail: plittle@mit.edu

A. Manto
e-mail: manto@mit.edu

W. McGee and M. Ponce de Leon (eds.), *Robotic Fabrication in Architecture, Art and Design 2014*, DOI: 10.1007/978-3-319-04663-1_1,
© Springer International Publishing Switzerland 2014

1 Introduction

Architecture has a long history of working with volumetric materials in a variable way. Only recently, as a result of the Industrial Revolution, has the building industry advocated for sheet materials and standardized building components (Clifford 2012). With greater attention paid to robotic fabrication in architecture, there has been a resurgence surrounding the topic of volumetric materials and their capacity to engage computational, algorithmic, free-form, or otherwise complex geometries that are not capable of being described through the Albertian ortho-graphic representations of architectural intent (Carpo 2011).

In recent years, designers have transferred Philibert de L'Orme's method and definition of *art du trait géométrique* (currently known as stereotomy) into the carving of large volumetric positives in expanded polystyrene (EPS) foam (Fallacara 2006; Feringa and Sondergaard 2014; Rippmann and Block 2011). Their projects, though working as an analog for stone construction, argued for the advantages of EPS for its regenerative abilities, lightweight, and machinability (Clifford and McGee 2011). A paper (Stavric and Kaftan 2012) expanded the carving of foam beyond the use of traditional linear cutting geometries into custom profiles. The use of custom profiles is not a new idea either, as it has been used historically for mold profiles in the method of plaster scraping as well as molding shapers for wood moldings and ornamental columns. A project (archolab.com/archives/40) recently applied the use of plaster scraping to robotic processes.

Recently, attention has been paid to the problem of creating complex molds to cast free-form geometries. Many projects have applied subtractive computer numerically controlled (CNC) technology to create custom formwork with high precision. This approach assumes the waste and non-repeatability of the molds in favor of further freedom in geometric creation. Another approach is to approxi-mate subtle curvature through the bending of sheet material against a super-structure (www.designtoproduction.ch/content/view/17/26/). This approach limits the global figure of the geometry to the maximum bending of the material in response to the Gaussian curvature. In a similar method, the use of a variable mold through pneumatics and actuators has been used to articulate a geometry via points across a malleable material in papers such as (Pronk et al. 2009; Raun et al. 2010). Gramazio and Kohler (dfab.arch.ethz.ch/web/e/forschung/164.html) have also demonstrated the advantage of re-usable materials to create serially variable molds. This method is not dissimilar to the use of earthwork as formwork for on-site or tilt-up concrete construction.

This chapter proposes a precise and rapid method for carving negative molds with a custom robotic hot-knife for highly variable free-form geometries without the material waste of typical subtractive machining approaches. It uses the column as an exercise to prototype this method and EPS foam as the carved material.

2 Digital Gothic

This chapter assumes a digital gothic approach as described by Ruskin (1960) in his text '*The Nature of Gothic*'. Ruskin describes the qualities of gothic as being determined by the maker and (his) methods of making. This approach can be conversely opposed to the classic approach of assuming the identical copies of a style that has been pre-determined by one 'thinker'. In establishing this dichotomy, one comprehends the history of division between thinker and maker, as well as the occasional alignment. With the advent of digital technologies in design and fabrication, our profession has found a reciprocal and harmonious relationship between the two. This chapter exercises one development in this reciprocity by generating sympathetic architecture (Carpo 2011; Spuybroek 2011).

3 Fluting and Bundling

The use of fluting or bundling (inversion) has been employed in the creation and subsequent ornamentation of columns for millennia, as demonstrated in Fig. 1. The earliest cataloged column types emerged in Egypt, Assyria, and Minoa 3000 BCE and were made by lashing reeds together at their ends caused them to cinch tightly to one another, and subsequently, bulge outward slightly at their midpoints. Later builders transferred these fluted geometries into stone construction. In his text '*Contrasts*', Pugin (1836) describes the classic period as "white, marbleized ghost of an essentially wooden architecture". Some of the earliest Egyptian stone columns even mimicked the bundled arrangement of reeds through the design of convex flutes. As column orders developed, fluting took on different roles. In the most recognizable column types, Doric, Ionic, and Corinthian, fluting developed by the Greeks as an analog to tool marks left in tree trunks as the bark was stripped away. They had concave, vertical flutes that were carved normal to the stone face. The global form also bulged at their midpoints (a technique known as entasis) referencing the previous reed columns and providing a visual effect that made columns appear more slim and elegant. Gothic builders bundled columns into piers in order to branch ribs above to create vaults. Later builders used fluting increasingly as bundling rhetoric like the twisting Solomonic columns of the baroque period that appear as two columns entwined. While this chapter does not advocate for the simple re-application of fluting as a stylistic choice, it does grapple with these issues due to a method of making constraint (see Sect. 5.1).

Fig. 1 Luxor Temple, Parthenon, Reims Cathedral, Hotel Neuschwanstein

4 Tooling

The scraping of tree trunks presented itself symbolically in ancient Greece via the flutes of columns. In some cases, tool marks were left as a vestige of the means of making. Figure 2 shows a room at *Les Baux-des-Provence* carved away from solid limestone and left with all of the tool marks from its excavation intact. These markings demonstrate a direct relationship between the global geometry, and the direction of the tooling. In other cases, tool marks are made much more deliberately and in an effort to efficiently create surface texture. Surfaces at *Hôtel Carnavalet* are chiseled in an inverted pyramid pattern and exhibit an efficient method of patterning, while the vermiculation carvings at the *Louvre* in Paris demonstrate a space filling patterning irrespective of direction (Clifford 2012). Ultimately, any carved surface bears the history of its tooling, though with rigid materials such as stone and wood, it is possible to further sand, polish, or finish a surface beyond the tooling. These processes are a balance between efficiency and resolution; however, as Pye (1968) argues in his text '*The Nature and Art of Workmanship*', nothing should be taken pride in that can be accomplished with some sandpaper.

In contemporary practice, tool-paths are commonly displaying themselves as the marks left by the machining of objects out of larger pieces of material as result of subtractive machining. In cases such as in *Commonwealth's* 'Lard Series' (http://www.commonwealth.nu/projects/61/lard), the overall form of the surface is quite deliberate. The parallel tool-paths are imprinted ever so slightly on the surface only as a trace of how the bench was made. Skylar Tibbit's 'Path Responsive Surface Milling' (http://sjet.us/phila_path_responsive_surface_milling.html) displays an entirely different attitude to the use of tooling in the production process. In his project, the tool-paths do not merely serve as a trace of production, but rather had a major role in the physical surface definition of the final object. Tool marks don't always have to imply movement over a surface. Rather, contemporary tools allow for the explicit control over tool entry, engage, and withdrawal motions.

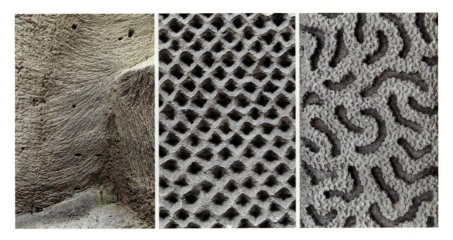

Fig. 2 *Les-Baux-des-Provence* Bouches-du-Rhone, France (*left*). *Hôtel Carnavalet* Paris, France (*middle*). *Louvre* Paris, France (*right*)

Avi Forman's work (http://www.flickr.com/photos/69844849@N00/5175767950) at the *Yale School of Architecture* shows an artifact formed through the use of thousands of individual drill points. The resulting surface emerges as a highly stippled combination of holistic form and nuanced individual tool-marking similar to the surface at *Hôtel Carnavalet*.

5 Methodology

The column prototype described in this chapter is fabricated with a robotically controlled hot-knife to carve negative geometries from EPS foam. These mold negatives are then sealed with a vacuum bag and used to cast unique Glass Fiber Reinforced Gypsum (GFRG) positives. This process involves the computing of the tool-paths, the rapid prototyping of potential forms, and the development of fabrication techniques used to make finished artifacts.

5.1 Hot-Knife Tooling and Geometric Constraints

Two hot-knifes were tested in the prototyping of this method. The first mounts an off-the-shelf hot-knife that is originally designed to be used by a human hand. It has a long bent handle that works well for a manual operation, but creates collision obstacles during robotic operations. The second version is a custom mounting that minimizes the handle and accommodates a larger blade. While this development does resolve the collision obstacle, both employed a semi-circular or 'J' blade

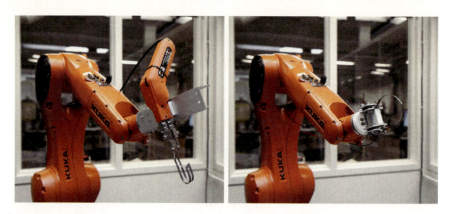

Fig. 3 A comparison of the two prototyped knives. Off-the-shelf craftsman hot-knife with a 1″ diameter 'J' blade (*left*). Custom hot-knife mounting with a 3.25″ diameter blade (*right*)

profile. The first is 1 in. in diameter and the second is 3.25 in. in diameter as shown in Fig. 3.

The hot-knife cuts through EPS foam by melting it at high temperature, creating both a minimum and a maximum step-over for the tool-path. Figure 4 demonstrates the maximum step over is a function of tool width; while minimum step over attempts to limit areas of re-melting from previous tool-paths. Typical CNC subtractive (milling) machining assumes there is a maximum, but no minimum. For this reason, milling can either express the tooling and enjoy the benefits of minimum passes, or increase the resolution of the step-over to greater approximate a smooth surface. This approach is dedicated to resolution, allowing the tool path to pass any way they are prescribed by the machinist.

While similarities can exist in appearance between these two methods, it is important to note that not all geometries can be translated from one to the other. Beyond the requirement of a minimum step-over, another major difference between milling and this proposal is that of direction. A milling tool-path has no direction because the blade is spinning. This also means a mill can turn a corner with zero radius, but the proposed method has a blade with a direction and therefore can only turn with a wide sweeping radius. The advantage this method has over milling is the ability to generate non-symmetric custom profiles.

Due to the constraints of the method, the approach demonstrated in this chapter is dedicated to the conscious decision of tooling location. This chapter does not advocate for the universal use of tool-path visibility, though this method of the hot-knife carving requires attention to this issue. Figure 5 shows a few of the tool-path strategies prototyped.

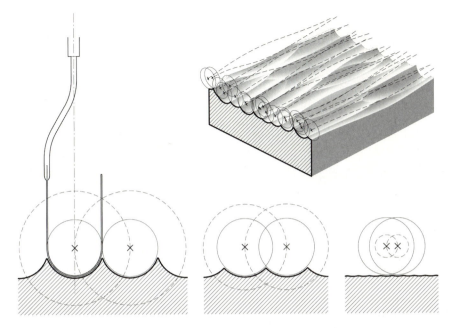

Fig. 4 A diagram describing the range between the minimum and the maximum step-over

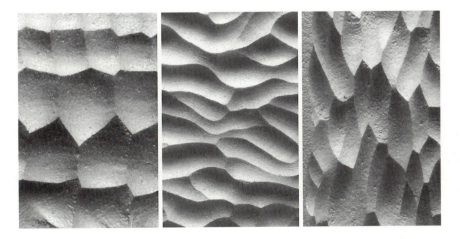

Fig. 5 Multiple tool-path prototypes

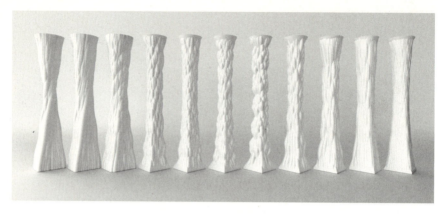

Fig. 6 Quarter sized rapid prototypes of serial variability

5.2 Digital Modeling and Computed Simulation

The column design is developed with a set of constraints such as the height (36 in. tall) and geometric transitions (from 6 in. square base to 6 in. diameter circle top) making this prototype comparatively proportionate to the Doric order. Figure 6 demonstrates a series of rapid prototypes. The forms are generated with curves that modulate normal to the surface of the column to create waves.

5.3 Materials

This prototype uses 2 lb/cu ft density Expanded Polystyrene (EPS) foam for the molds. The Glass Fiber Reinforced Gypsum (GFRG) used to cast into this mold is comprised of Hydrocal, a white gypsum-based cement that has higher strength (up to 5,000 psi) and quicker setting time than typical plaster. It also uses Chopped Strand Glass Fibers (CSGF) as reinforcement to increase mechanical strength of the Hydrocal. 1/4″ long fibers are chosen due to the high degree of surface resolution necessary in the final casting. The two part cast is then filled with Expanding Water Blown Urethane Foam, a two-part castable rigid foam that expands many times its original volume when mixed together.

5.4 Multi-Axis Robotic Carving

Multi-axis robots are a type of programmable machine with multiple rotary joints. The tests undertaken in this chapter are conducted on a KUKA KR6 R700 6 axis robot. The maximum reach of the extended robot arm from its base position is

Fig. 7 Robot and hot-knife assembly carving EPS foam

roughly 28 in., but the necessary positioning of the arm for certain geometries often means the reach is less. The $12'' \times 36''$ block of EPS foam pushes the reach of the machine and necessitates a rotated stock orientation to ensure tool-paths stay within reach. Figure 7 shows the first knife fitted with a semi-circular blade mounted at the end of an aluminum extension holding the knife. Curves are extracted from the digital model and subdivided with 'normal planes', which the robot uses to correctly align and guide the hot-knife through the material.

5.5 Casting/Mold Making

The design of this column requires two mold negatives used to cast two positive column halves that are brought together to form a single column. The required yet added texture of the tool-path geometries in the molds made releasing the casts difficult.

A number of demolding strategies were tested, from a laborious process of applying and sanding joint compound over the surface to mold wax, vegetable oil, and even water based spray release agents. Occasionally poor or uneven application can create difficulty in demolding. When these stubborn areas do not release, solvents are applied to melt the EPS foam away from the GFRG cast.

The preferred method for demolding involves stretching a thin latex sheet over the mold and applying a vacuum, effectively sucking the latex to the face of the mold. This technique creates a sealed surface into which plaster can be cast and serves as an excellent mold release system. The first tests utilize a relatively thick latex sheet ($0.07''$ thick), which creates an alluringly smooth surface on the casting, but also deformed and rounded the edges of the mold to a degree that was unacceptable. A thinner sheet of latex ($0.02''$ thick) is used next to achieve a higher resolution finish. An unintended by-product is the change in finish quality from a highly polished and shiny surface to smooth, but textured as seen in Fig. 8.

Fig. 8 Comparative resolution between 0.07″ latex sheet (*left*) and 0.02″ latex sheet (*right*)

This method is rapid, efficient, and durable. The resulting surface finish is also highly controllable given the variables of foam density and latex thickness.

5.6 Assembly and Finishing

The two plaster casts are sanded along their common joining faces until they are both planer. The halves are temporarily held together with tape while two-part expandable urethane foam is poured in the hollow cavity. The foam expands and hardens, permanently joined both halves together.

6 Analysis

A number of questions, concerns, and future research goals have been established. The hot-knife proves to be an expedient way of creating complex surface features in EPS foam, but its use is met with certain difficulties. Finding the center point of hot-knife blade and creating proper alignment on the robot is difficult due to the flexible nature of the blade, and its lack of clear registration points. Also, blade temperature fluctuates depending on the depth and time of the cut, which led to an excess of melted foam when too hot, and tear-out of foam when too cool. We found these difficulties easy to overcome and further research will engage a feedback loop with a temperature sensor to speed up or slow down the carving path. We also see value in the creation of custom blade profiles.

This method is not dedicated exclusively to columns. Future work will address the gothic method of branching and bundling as seen in fan vaults and column piers. Gouging has the potential to be re-directed, multiplied, or extended in response to transitions or anomalous conditions in free-form geometries. The 'macro-mark' contains another advantage particularly aligned with robotics. These large gouges are not aligned to resolution, and therefore the possibility of translating large material within the reach of the arm can be accomplished without the requirement of perfect indexing. This strategy is akin to the feathering of hair, where it is not clear where one set of paths start and the others begin.

Though this prototype employed GFRG and expanding urethane foam, there is nothing about this method that is strictly dedicated to these materials. Future work will expand this material pallet with the similar mold creation method.

7 Conclusion

This chapter presents a process for producing mass-customizable formwork for free-form geometries without material waste. It successfully demonstrates the practicality and potential by rapidly carving variable molds to cast unique volumetric architectural elements. The foam molds as well as the releasing process used to cast these elements are 100 % recyclable.

While this chapter presents a column as prototype, the method is in no way dedicated to this element typology. We understand the column as a placeholder for a variety of architectural forms to come. For this reason, we do not test the column for its structural capacity; rather, we test it against precision, ability to cast, and ease of fabrication. This process produces a form that is highly unique to its methods of making. We learned the limitations and constraints of the method are directly aligned with a long history of volumetric carving. This re-insertion of past knowledge into contemporary methods results in a new language that has roots in ancient traditions.

We see robotics as translating gothic craft traditions into a digital environment, full of feedback and variability. While this method could be produced roughly with a number of non-robotic controllable devices, we see it uniquely aligned with robotics for their inclination toward volumetric processes and system feedback. This method is also inherently lightweight and lacks the precision one would expect from CNC subtractive machining. Where it lacks in this precision, it makes up for in efficiency of scale. One must keep this false assumption of precision in consideration when developing a design strategy as the method is embedded with a few potentials for aligning units we are excited to test in future work. In prototyping this hypothesis, we learned a number of "fuzzy" variables that could be better informed by sensing and system feedback, for instance, the over-melt of previous passes and the variation in knife temperature versus the feed.

We see no limit to the forms or elements this method can produce, as it is not dedicated to a scale, material, or style. It is dedicated to an efficiency of scale, a limitation of waste, and the digital desire for variation. This chapter opens the doors of variation to methods previously relegated to the re-production of identical forms.

Acknowledgments This chapter is part of a research workshop at the *Massachusetts Institute of Technology* titled 'Volumetric Robotics' and led by Brandon Clifford. This course is dedicated to translating volumetric methods of making into contemporary digital culture. Teaching assistance by Trygve Wastvedt and Bobby White. Fabrication and facility support by Justin Lavallee and Larry Sass. Geometries were generated in *Grasshopper* (grasshopper3d.com), a parametric plugin developed by David Rutten for *Rhinoceros* (rhino3d.com), a program developed by Robert McNeil. Robotic control was generated via *Hal* (hal.thibaultschwartz.com), a program developed by Thibault Schwartz and *SuperMatterTools*, a program developed by Wes McGee of *Matter Design* (matterdesignstudio.com) and Dave Pigram of *Supermanoeuvre* (supermanoeuvre.com).

References

Carpo M (2011) The alphabet and the algorithm (writing architecture). The MIT Press, Cambridge

Clifford B (2012) Volume: bringing surface into question. SOM Foundation final report

Clifford B, McGee W (2011) Matter and making: periscope foam tower. In: Glynn R, Sheil B (eds) Fabricate: making digital architecture. Riverside Architectural Press, London, pp 76–79

Fallacara G (2006) Digital stereotomy and topological transformations: reasoning about shape building. In: Proceedings of the second international congress on construction history, vol 1, pp 1075–1092

Feringa J, Sondergaard A (2014) Fabricating architectural volume: stereotomic investigations in robotic craft. In: Gramazio F, Kohler M, Langenberg S (eds) Fabricate: negotiating design & making. gta Verlag, ETH Zurich, pp 76–83

Pronk A, Rooy I, Schinkel P (2009) Double-curved surfaces using a membrane mould. In: Proceedings of the international association for shell and spatial structures (IASS) symposium 2009, Valencia, Spain

Pugin A (1836) Contrasts: or, a parallel between the noble edifices of the fourteenth and fifteenth centuries and similar buildings of the present day. Shewing the present decay of taste. Accompanied by appropriate text, Salisbury

Pye D (1968) The nature and art of workmanship. Cambridge University Press, London

Raun C, Kristensen M, Kirkegaard P (2010) Flexible mould for precast concrete elements. In: Proceedings of the international association for shell and spatial structures (IASS) symposium 2010, Shanghai, China

Rippmann M, Block P (2011) Digital stereotomy: voussoir geometry for freeform masonry-like vaults informed by structural and fabrication constraints. In: Proceedings of the IABSE-IASS symposium 2011, London, UK

Ruskin J (1960) The nature of Gothic. In: Links JG (ed) The stones of Venice. Da Capo Press, New York

Spuybroek L (2011) The sympathy of things: ruskin and the ecology of design. V2_Publishing, Rotterdam

Stavric M, Kaftan M (2012) Robotic fabrication of modular formwork for non-standard concrete structures. In: Achten H, Pavlicek J, Hulin J, Matejdan D (eds) Digital physicality—Proceedings of the 30th eCAADe conference, vol 2 12(14), pp 431–437

Bandsawn Bands

Feature-Based Design and Fabrication of Nested Freeform Surfaces in Wood

Ryan Luke Johns and Nicholas Foley

Abstract While the rising trend of research in robotic fabrication has furthered the development of parametric or mass-customization concepts in architecture, the majority of these projects are still cut or assembled from standardized blocks of material. Although the use of nonstandard, 'found' components provides an additional layer of complexity and constraint to the design/fabrication process, it can compensate for these challenges by enabling more sustainable material practices and the production of unique objects that cannot be reproduced. In this chapter, we illustrate a materially efficient technique for designing and fabricating freeform surfaces within the constraints of irregular wood flitches. The process utilizes a robotically operated bandsaw to cut a series of curved strips which, when rotated and laminated, can approximate doubly-curved and digitally defined geometry. By delimiting the design space by both the 'machinic morphospace' (Menges in Rob|Arch 2012: Robotic Fabrication in Architecture, Art and Industrial Design, Springer, Vienna, pp. 28–47, 2012) of the fabrication technique and the naturally defined curvatures and constraints of the flitch, the customized control software and machining processes confer the capabilities of digital fabrication onto materially tailored and operator-informed woodcraft.

Keywords Robotic bandsaw · Freeform surface · Timber construction · Parametric design · Minimal waste

R. L. Johns (✉)
Greyshed and Princeton University, Princeton, NJ, USA
e-mail: ryan@gshed.com

N. Foley
Greyshed, Princeton, NJ, USA
e-mail: nick@gshed.com

W. McGee and M. Ponce de Leon (eds.), *Robotic Fabrication in Architecture, Art and Design 2014*, DOI: 10.1007/978-3-319-04663-1_2,

1 Introduction

Despite the capabilities for customized fabrication offered by digital tools, the vast majority of parametric design-fabrication exercises remain confined to the parameters of standard, industrially produced components. Whether by creating "highly informed" (Bonwetsch et al. 2006) geometry from an additive assemblage of standardized parts, or from a series of subtractive operations upon them, the variable control provided by the generic component (blanks, bars, bricks and sheet stock) is commonplace. While the economy of scale renders this practice practical for many applications, its shortcomings reveal the necessity to explore alternate models of production which engage the complexity of found materials. Rather than transferring material, for example, from a curved tree into dimensional lumber which is then re-machined into curvilinear digitally designed geometry, we take the tree as the starting point for design and move directly to digital fabrication. This leap in the production sequence enables more sustainable material efficiency while simultaneously conferring the natural aesthetic advantage of "beauty's found geometries" (Enns 2010) (Fig. 1).

In this chapter, we document a technique for designing and fabricating freeform surfaces from live-edged wood flitches. The process utilizes a robotically operated bandsaw to cut a series of curved strips which, when rotated and laminated, can approximate doubly-curved and digitally defined geometry. The thin kerf of the bandsaw blade allows for a tight nesting of finished surfaces, which, through a close relationship between available material and designed geometry, affords practically zero-waste when compared to Computer Numeric Control (CNC) contour milling. This coordination of non-standard material geometries, 3D scanning, parametric algorithms, designer input and robotic control serves to bolster the role of feature-based material intuition in design and "digital craft" (Johns 2014; McCullough 1996).

2 Related Work

The practice of linking 3D surface scanning, tomography, and feature recognition to more efficiently process logs has been present in the lumber industry for quite some time (Conners et al. 1983). This process, however, is generally used as a means to work around defects and irregularities to achieve a higher yield and grade of standard dimensional lumber. In contrast, this project is specifically interested in reading specific and non-standard geometries *from* the irregularities rather than an attempt to embed rectilinear objects within them.

There are a variety of projects which focus upon geometrical nesting of discrete industrial design objects, from Tom Pawlofsky and Tibor Weissmahr's "7xStool" (http://www.kkaarrlls.com/index.php?feature=editions,7Xstool), to Karim Rashid's "Matryoshkarim" (http://www.detail.de/daily/matryoshkarim-von-karim-rashid-6307/),

Fig. 1 Robotic bandsawing of irregular flitch

and as far back as Albert Decker's 1925 patent for "Nesting Furniture" (Decker 1928) (Fig. 2). These types of projects generally create a family of objects which interlock into a standard bounding box (or live-edged, but relatively standard log, in the case of the 7xStool) for simplicity of fabrication, material efficiency or compactness for storage. This object-based approach to packing has powerful implications, but in practice, generally results in several objects with a high performance value (a chair or table) and one or several "remainder" objects with ambiguous functions (a paperweight or footrest). Rather than attempting to solve the difficult problem of packing a variety of discrete objects within a volume, this research focuses on the reorganization and assembly of a found material condition into a single and continuous entity.

Digital fabrication techniques for more efficient freeform surface production have been explored in the CNC-milled "Zero/Fold Screen" by Matsys (http://matsysdesign.com/2010/02/28/zerofold-screen/), and made more viable for continuous surface fabrication with the capabilities of 7-axis swarf milling to produce a "single cut…finished surface" (Brell-Cokcan et al. 2009). The comparatively wide cut path and short length of the CNC router, however, prohibits the same degree of surface mating made possible with the bandsaw. For an examination of additional thin kerf abrasive-wire techniques in robotic fabrication, refer to "Processes for an Architecture of Volume" (McGee et al. 2012).

Fig. 2 From *left* Pawlofsky and Weissmahr's "7xStool," Rashid's "Matryoshkarim," Decker's "Nesting Furniture"

3 Algorithm and Implementation

3.1 Freeform Volume Nesting

The nesting technique is based upon the premise that the bottom face of each sawn band is a copy of its neighbor's upper face. The design control software is generated using both *Grasshopper* and *Python* within the *Rhino* modelling software. It operates as a multiple stage process in the following manner:

Flitch-Fitted Surface Generation:

1. The process for creating nested freeform volumes within the confines of a flitch[1] of thickness (t_{flitch}) with volume (V_{flitch}) begins with the digitally modelled live edge surfaces (s_1 and s_2) of the 3D scanned flitch (Fig. 3-1). In this implementation, the scan is achieved using either the Kinect or the robotic manipulator as a digitizer.
2. The freeform surface (S_f) is parametrically linked to the control variables of the available flitch. Specifically, its length must be equal to the length of the flitch, and its width (W_{sf}) will be a factor of the thickness of the flitch multiplied by the number of bands (Eq. 1) (Fig. 3-2).

$$W_{sf} = t_{flitch} \times n_{bands} \tag{1}$$

The edge curves along the length of the surface must generally follow the curvature of s_1 and s_2, and, for surface continuity, will ideally also maintain tangency.
3. The above conditions provide an infinite subset of permutations which must be navigated by the designer to produce the desired surface, S_f (Fig. 3-3).

---

[1] Note that it is possible to generalize this process to fabricate surfaces which do not operate within the confines of a flitch, by proceeding directly to step 4.

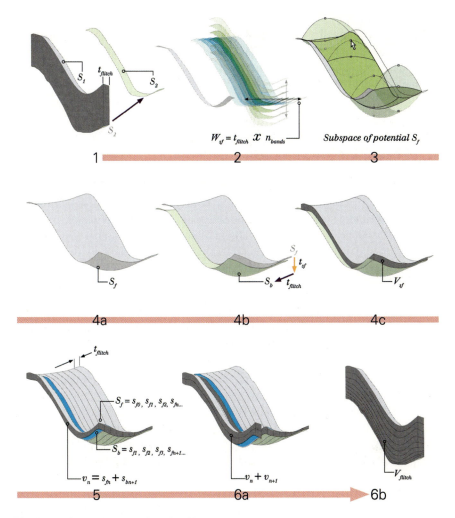

Fig. 3 Freeform volume nesting algorithm

Freeform Volume Nesting:

4. The designed surface S_f is translated horizontally by t_{flitch}, as well as translated vertically by t_{sf}, an amount that will determine the thickness of the final volume created (V_{sf}). The resultant surface is S_b (a translated duplicate of S_f), and the enclosed region between S_f and S_b becomes the boundary of the final volume created (V_{sf}) (Fig. 3-4).

5. The surfaces of S_f and S_b are now divided along their width into a number (n_{bands}) of equally spaced sections (s_{fn}) of thickness t_{flitch}. Because of the translation used to create S_b, any subsurface of S_b (s_{fn}) is the same as the next subsurface of S_f (s_{fn+1}) (Fig. 3-5).

6. The sections of S_f and S_b are now paired vertically, being treated as the upper and lower bounds of a sub-volume (v) of V_{sf}. Each v can be nested beneath the next, collapsing V_{sf} into a fully dense planar volume (Fig. 3-6).
7. The precise value of t_{sf} is determined using a simple binary search algorithm which seeks to set V_{sf} equal to V_{flitch}.

3.2 Initial Studies

As custom end effectors and irregular material conditions can quickly complicate the prototyping process, our initial studies utilized standard dimensional lumber and unmodified tools in order to ease investigation. In these experiments, the robot is holding the work object ($2 \times 12''$ nominal lumber) with a simple toggle clamp end effector, and manipulating the wood through a standard bandsaw (Fig. 4b) and a wall mounted benchtop grinder which has been outfitted with drill chucks such that it may serve as a double sided drill (Fig. 4a). This setup has the advantage of being able to operate as one continuous program, without the need to manually change end effectors between the dowel drilling operation and the sawing operation. This prototypical process, however, has obvious limitations of board length and maneuverability, as the robot must make increasingly large movements when making cuts that are further from its adaptor plate.

These initial physical experiments revealed numerous shortcomings, which were then reprogrammed into the system. For example, we prolonged the life of blades and shortened the duration the cut sequence by programming the robot's cut speed as a value proportionate to the twist and curvature of the cut. Similarly, we added an implementation of the travelling salesman problem (http://www.psychicorigami.com/2007/04/17/) to the dowel drilling sequence in order to decrease cycle times.

As these prototypes were not cut from an irregular flitch, their design constraints are limited by the width of the dimensional lumber and the curvature limitations of the selected bandsaw blade (~ 50 mm minimum cut radius). The first prototype emulates a possible found-flitch geometry, while the second works in the reverse direction: beginning with a desired surface and working backwards (Fig. 5). In this instance, we use a Kinect scan of an individual falling in the seated position as the guiding geometry of the surface.

While these initial studies were instrumental in fine tuning the control parameters and in recognizing physical limitations of the tools, they quickly reemphasized the shortcomings of attempting to nest curvilinear designs within rectilinear volumes.

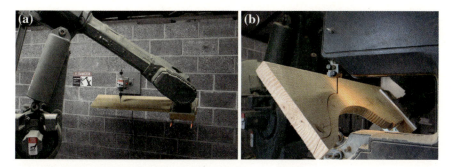

Fig. 4 From *left* **a** Double ended, wall-mounted dowelling drill. **b** Robotic operation of standard bandsaw

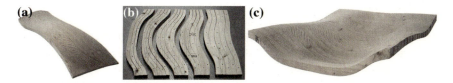

Fig. 5 **a** Freeform surface emulating flitch geometry. **b** Nested bands separated by board origin. **c** Laminated surface

3.3 Live-Edged Prototype

In order to optimize the setup to facilitate irregular flitches of various shapes and sizes, the process was inverted such that the flitch is stationary and the robot is holding the tools. For the dowelling end effector, we use a router and closed-loop speed controller to allow us to maintain the high torque at the low RPM's suitable for the 3/16″ dowelling drillbit (http://www.vhipe.com/product-private/SuperPID-Home.htm). The bandsaw end effector is, quite literally, the standard 12″ bandsaw removed from its base, reinforced with a welded steel frame, and mounted onto the robot.

In order to accommodate a variety of possible flitches, we oriented a wooden column within the robot's reach envelope, into which the robot drilled an array of holes for attaching mounting hardware. By using the robot to construct the machining jig within which it operates, we are ensured that the flitch can be oriented precisely within the coordinate system of the robot with little effort in calibration. The robot also drills a matching array of holes into one live edge of the flitch, strategically placing the holes such that they are in the thickest areas of the designed band and to a depth less than the thickness of the band in that region. Wood-screw-threaded studs are then manually inserted into these holes, and become the mechanism for mounting the flitch to the column. In this way, we avoid the use of bulky clamps through which the saw cannot pass, and are able to

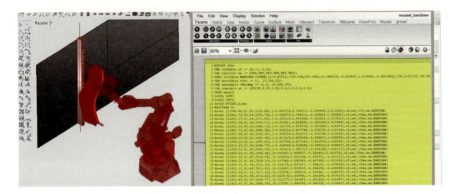

Fig. 6 RAPID code for bandsaw process generated and previewed using the Mussel plugin for Grasshopper

process the entirety of the surface in one operation. These studs can also be later repurposed in the completed surface as connection points to a support structure (in example, the legs of a chair).

The selected flitch for this prototype was chosen with a close consideration of the outlined algorithmic process, recognizing that its edge curvature implied the potential of a chaise longue. Following the design of the surface, the generated drill and cut paths were converted to *RAPID* (ABB's robotic programming language) using the open-source *Grasshopper* plugin *Mussel* (http://www.grasshopper3d.com/group/mussel) and fabricated with the IRB-6400 robot with S4C controller (Fig. 6).

4 Discussion

4.1 Process Benefits

The process allows for a high material efficiency due both to the machining process (the bandsaw has the smallest possible kerf of any mechanical woodcutting method) and the designs being parametrically customized to the workpiece in order to eliminate waste. As with swarf milling (Brell-Cokcan et al. 2009), there is also a high process efficiency, as each cut operation forms two finished surfaces: the bottom of one piece and the top of the next. Similarly, the 6-axis control of the robotic manipulator allows the fabrication of cuts which are not only curved in two dimensions, but (by "twisting" the bandsaw blade) can create three dimensional ruled geometries which, when combined, can smoothly approximate a doubly-curved surface. This allows a much cleaner surface resolution than the traditional terraced geometries arising from laminated 2.5-axis CNC operations.

Fig. 7 *Left* sample flitch, with nested cuts and dowel holes. *Right* in-progress assembly

The technique also holds an interesting position as one of few woodworking techniques which are explicitly not "subtractive" (Kolarevic 2003), but *transformative* (Fig. 7). By parametrically generating the design as a reconfiguration of the workpiece, the process not only enables use of any individual material piece more efficiently, but enables use of more of the complete tree. This form of 'nose-to-tail consumption' bears more resemblance to highly efficient and direct-to-consumer automated butchery (Loeffen and Purnell 2006) (http://www.scott.co.nz/meat-processing/lamb/automated-boning-room-systems) than to many traditional sawmill operations. Sections which would have previously only yielded small amounts of dimensional lumber due to natural curvature or inclusions can be utilized in their entirety because those features are integrated into the design process. This allows the direct use of widely available and economical "wind fall" lumber, which is frequently not suitable for traditional applications.

4.2 Design Benefits

By fabricating the surface as a reconfiguration of a continuous slab, the final product visually exhibits both the narrative of the source material and the production technique, as traced through repetition and continuities in grain pattern, knots, and imperfections (Fig. 8). This narrative can also be expanded further through the use of multiple sections from one tree, either creating a larger continuous surface (by joining the corresponding live edges of subsurfaces) or a range of related product families. A single "design" can take on subtle permutations which adapt based on the particular tree section chosen, while being united with other pieces from the same tree in general design, woodgrain, and bark conditions. The close association of found geometry and produced artifact ensure not only that each object will be uniquely tailored, but that it also cannot be reproduced. This statement does not hold true for projects which utilize industrially produced source materials (plywood, MDF, EPS foam, etc.): provided that the source code remains, any individually tailored geometry can be replicated. The role of grain, while promoting a specific aesthetic agenda, simultaneously exhibits optimal structural properties: by informing the designed form with the geometry of the live edges, the

Fig. 8 Detail of prototype in cherry

natural anisotropy of the woodgrain is aligned with the cutting operations and final part output in turn.

This more materially efficient process confers the advantage that rarer and more expensive woods can be used in complex freeform surfaces which would otherwise need to be CNC-milled from a block of material many times larger (Fig. 9). Not only a cost benefit, this also eliminates the design constraint of sourcing a large block of material: as most CNC-milled wood projects are presently made with Ash, Birch, or Fir because these materials are relatively inexpensive and available in large dimensions.

4.3 Improvements

The current state of the project has room for an array of potential improvements, in both the design process and its physical execution. Presently, the assembly system relies on the bands being joined with dowels and wood glue. While this process is generally straightforward, it becomes less feasible with large surface deformations. Clamping such geometries without a jig which has been fabricated specifically to the piece can be time consuming (if not impossible). However, the waste associated with the fabrication of such a jig would override many of the efficiencies of the project. A mechanized jig which allowed for a wide array of configurations could be developed, or potentially, some of the dowels holes could be replaced

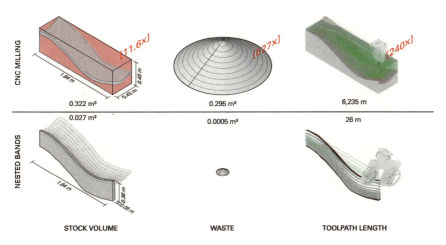

Fig. 9 Comparison between waste, volume, and machining requirements for fabricating a sample surface with CNC milling versus flitch nesting

Fig. 10 Doubly-curved chaise longue surface made from irregular flitch

with through-holes which allow a steel cable to pass through the bands in a post-tensioning system. Each band could also simply be screwed to the previous band. The latter two alternatives, however, have the downside of somewhat limiting the minimum possible thickness of the bands.

The prototypes presented in this chapter rely on as-is lumber for which we have little control of thickness, planarity, or cut plane. Ideally, the process could develop further through a close coordination with the sawmill so that *all* cuts are both digitally, and designer, informed. We imagine a potential future where a library of approximate desired geometries (i.e. a variety of seat shapes) can be cross referenced with the scanned tree using a technique of "fuzzy correspondences" (Kim et al. 2012) in order to isolate potentially fruitful irregularities from the stock which is more suitable for dimensional lumber (Fig. 10).

Fig. 11 Chaise longue assembled from nested bands

5 Conclusion

In this chapter, we present a materially efficient technique for the feature-based design and fabrication of freeform surfaces within the constraints of an irregular wood flitch. The design is therefore not determined solely by the "machinic morphospace" (Menges 2012), but more importantly by the found and organically defined material morphospace, delimited by the edge conditions and dimensions of the flitch and the constraints of the outlined algorithm. The process enables a unique "cultural performance" (Oesterle 2009) to be pulled from otherwise unusable lumber through a process of materially tailored design. Rather than simply introducing a procedure which is materially efficient, the process is significant in that it makes it necessary to strike a balance between the beauty of the naturally formed, as-is material object and the geometrical demands of the designer, a relationship that is much more prominent in traditional handicraft than in digital fabrication (Fig. 11).

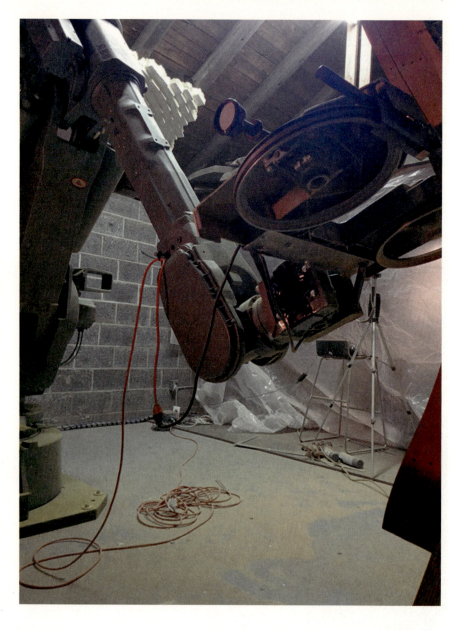

Acknowledgments Many thanks to Axel Kilian for his support during the early phases of this research.

References

Bonwetsch T, Gramazio F, Kohler M (2006) The informed wall: applying additive digital fabrication techniques on architecture. In: Proceedings of the 25th annual conference of the association for computer-aided design in architecture, Louisville, Kentucky, pp 489–495

Brell-Cokcan et al (2009) Digital design to digital production: flank milling with a 7-axis CNC-milling robot and parametric design. In: Proceedings of the eCAADe conference, Istanbul, Turkey, pp 323–330

Conners et al (1983) Identifying and locating surface defects in wood: part of an automated lumber processing system. IEEE Trans Pattern Anal Mach Intell 5(6):573–583

Decker AK (1928) Nesting furniture. US Patent 1660119A, Filed 24 Oct 1925, Seattle, WA

Enns J (2010) Beauty's found geometries. M. Arch thesis, Princeton University (Advisor: Prof. Axel Kilian)

Johns R (2014) Augmented materiality: modelling with material indeterminacy. In: Gramazio F, Kohler M, Langenberg S (eds) Fabricate. gta Verlag, Zurich, pp 216–223

Kim V, Li W, Mitra N, DiVerdi S, Funkhouser T (2012) Exploring collections of 3D models using fuzzy correspondences. ACM Trans Graph (Proc SIGGRAPH)

Kolarevic B (ed) (2003) Architecture in the digital age: design and manufacturing. Spon Press, New York

Loeffen M, Purnell G (2006) Automation for the modern slaughterhouse. In: Nollet L, Toldrá F (eds) Advanced technologies for meat processing. CRC/Taylor & Francis, Boca Raton, pp 43–71

McCullough M (1996) Abstracting craft: the practiced digital hand. MIT Press, Massachusetts

McGee A, Feringa J, Søndergaard A (2012) Processes for an architecture of volume: robotic hot wire cutting. In: RoblArch 2012: robotic fabrication in architecture, art and industrial design (2013). Springer, Vienna, pp 28–47

Menges A (2012) Morphospaces of robotic fabrication: from theoretical morphology to design computation and digital fabrication in architecture. In: RoblArch 2012: robotic fabrication in architecture, art and industrial design (2013). Springer, Vienna, pp 28–47

Oesterle S (2009) Cultural performance in robotic timber construction. In: Proceedings of the 29th annual conference of the association for computer aided design in architecture (ACADIA), Chicago, pp 194–200

An Investigation of Robotic Incremental Sheet Metal Forming as a Method for Prototyping Parametric Architectural Skins

Ammar Kalo and Michael Jake Newsum

Abstract Given its relative low-cost, speed and versatility, incremental sheet metal forming promises to introduce new ways in which architectural sheet metal cladding components are designed, prototyped and fabricated. Expanding on research done on this fabrication method, this work aims to study Single Point Incremental Forming (SPIF) and Double-Sided Incremental Forming (DSIF) as a viable option to produce highly customized, performative architectural skins. Utilizing the reconfigurable potentials of robotic arms' versatile tooling, multi-axial positioning, and simultaneous programming, new methods are integrated into the forming process for structuring, verifying and articulating parametric parts.

Keywords Single point incremental forming · Robotic fabrication · Metal forming · Parametric design · Architectural skins

1 Introduction

Since the discovery of cast iron, metals have radically transformed the way in which architecture is built and conceived. Metal preformed under its structural capacity and physical characteristics up until the early 20th century when a shift from structural functionalism to aesthetic surfacing occurred. Once sheet metal became a staple material of a larger architectural material palette, ways to manipulate its surface became of extreme interest to architects and designers. Two dimensional

A. Kalo (✉)
University of Michigan, Ann Arbor, MI, USA
e-mail: amarkalo@umich.edu

M. J. Newsum
Southern California Institute of Architecture, Los Angeles, CA, USA
e-mail: jake_newsum@sciarc.edu

W. McGee and M. Ponce de Leon (eds.), *Robotic Fabrication in Architecture, Art and Design 2014*, DOI: 10.1007/978-3-319-04663-1_3,
© Springer International Publishing Switzerland 2014

fabrication processes have proven their efficiency in mass customization, but they limit the designed geometries to flat or developable surface topologies.

In light of these recent production methods, incremental sheet metal forming promises to be an extremely viable option to produce articulated and varied double curved surfaces for architectural cladding systems. Current incremental metal forming processes cannot compete with the speed in which stamped facade panels are produced, nor can they compete with their extreme precision and low tolerances. However, they do provide opportunities to prototype designs and produce highly variable and cost effective components. When compared with surfaces composed of multiple joined developable surfaces, the double curved geometries possible with incremental forming yield inherently self-structured panels.

The research investigates forming techniques using single point incremental forming (SPIF) and dual side incremental forming (DSIF), but also a closed loop design and fabrication process where the properties of material, the tools used and geometry are intrinsically linked. Understanding both the limits and potential of the process is essential. Through a process that encompasses digital geometric design, toolpath generation, and validation methods, designers can parametrically explore cost effective options and quickly prototype them. The different areas of exploration are always bracketed within an architectural context and viewed through the lens of contemporary methods of fabrication.

1.1 Relevance of Incremental Forming in Architecture

In 2001, Asymptote architects designed Hydra Pier which is an information center that was the first "blobby" building in the Netherlands. Its complex surface produced double curved panels which were formed using explosion forming. This technical innovation at the time was, and still is, a very tedious process of producing multiple positive and negative molds of different materials for each unique panel (Eekhout 2008). While the Hydra Pier project demanded multiple panels to be shaped uniquely in order to achieve a unified whole, the metal façade of High Line 23 designed by Neil Denari is an assembly of the same stamped panel that assembles into a tile-able pattern. Stamping sheet metal is usually associated with the automobile industry, but in this instance the architect employed this technology to gain surface variation (Simmons 2008). Incremental sheet metal forming in the near future could allow architects and designers to fabricate variable complex facade systems on large scale, whether it's a 'wrinkled' skin like Frank Gehry's New York residential tower or complex geometries with undercuts which cannot be manufactured with conventional stamping and only possible with the multi-axis freedom of an industrial robot.

Incremental forming is a currently studied as a feasible replacement for stamping, at least in the prototyping phase. The Ford Motor Company has been awarded $7.04 million by the U.S. Department of Energy to develop this technology further into a robust mass production process (http://www.at.ford.com/news/cn/Pages/

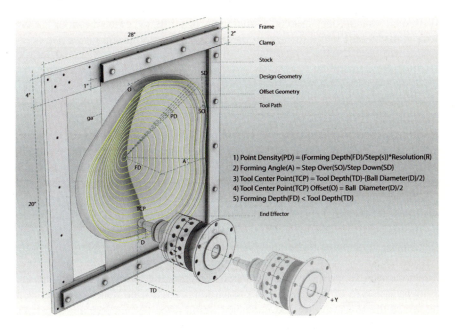

Design Geometry

Offset Geometry

Tool Path

1) Point Density(PD) = (Forming Depth(FD)/Step(s))*Resolution(R)
2) Forming Angle(A) = Step Over(SO)/Step Down(SD)
3) Tool Center Point(TCP) = Tool Depth(TD)-(Ball Diameter(D)/2)
4) Tool Center Point(TCP) Offset(O) = Ball Diameter(D)/2
5) Forming Depth(FD) < Tool Depth(TD)

Fig. 1 Single point incremental forming with a custom forming end effector

FordDevelopsAdvancedTechnologytoRevolutionizePrototypingPersonalization LowVolumeProduction.aspx). When comparing the two processes, stamping and incremental forming, the former promises better energy savings when compared one to one in an established setup (Ingarao et al. 2012). However, for highly customized panels incremental forming becomes more efficient as it doesn't require the production of unique dies for each panel. With the process' ability to rapidly prototype and produce full-scale formed metal panels, incremental forming begs for a thorough investigation in other disciplines such as architecture and the building construction industry.

1.2 Process Overview

SPIF and DSIF are methods by which flat sheets of metal are formed creating three dimensional components with variable surface curvatures. This process of forming, unlike stamping or molding, is able to produce variety without the need of formworks or dies. A spherical forming tool, mounted on a robotic arm in this case, moves along a preprogrammed path that is generated from a 3D model. The path starts from the perimeter of the designed part and continuously pushes against the surface of the sheet metal until it reaches the center of the deepest concavity in the design (Fig. 1). Additional tools and processes have been incorporated with

SPIF, such as additional structuring tool paths or DSIF, to reduce the stock's creep from global forces while increasing the part's fidelity to the digital model.

1.3 Recent Work

Work described in this chapter is a continuation of a larger research trajectory aimed at developing the necessary tools and techniques to produce discrete parametrically designed prototypes of building skins. The work also largely builds on previous efforts that focus on computer modeling methods, parametric tool path generation, forming practices, material testing, part validation, and aggregation strategies (Kalo et al. 2014). Several materials were tested to find the limitations of the forming process and for viability. The sampling of sheet metal included cold rolled steel, aluminum, copper and brass. Cold rolled steel was used in most of the forming tests because of its high ductility and strength, which allowed for forming deeper parts. Also, previous efforts looked at integrating forming inaccuracies and springback as part of the design and fabrication process itself. Once trimmed from the rest of the stock, the formed geometry would often deform into a new relaxed shape. This release of internal forces is called springback. The deeper a panel is formed, the more it would become rigid and resist this deformation. Several design strategies, described later in the chapter, were utilized to stabilize the components pre-trimming.

2 Methods and Techniques: Single Point Incremental Forming

With SPIF, the forming limitations with one tool are that the designed geometry concavity may only be in the forming direction and the accuracy of the formed part will decrease with proximity to the perimeter. A parametric tool path allows the designer to dynamically set the resolution in terms of radial divisions and number of revolutions using a custom script written within the 3D modeling program. Bezier graph controls in the script can locally increase or reduce the resolution to compensate for uneven surface variation as well as a means of overcoming shallow forming slopes.

Most researchers, concerned with high precision forming, have used very high resolution settings in order to produce extremely refined parts. Additionally, the size of the forming stylus, or tool, plays an important role in the final surface finish. For the purposes of this research, since industrial grade tolerances and manufacturing precision is not the main goal, a medium 15 mm custom forming tool has been used. The surface 'roughness' could be predicted with percentage

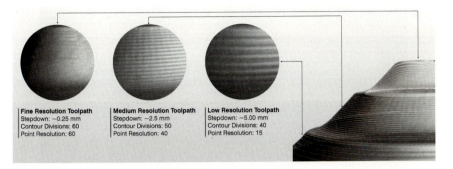

Fig. 2 Samples of the varied resolutions that can be achieved

differences that don't exceed 10 %, however, the path resolution here is treated as a design parameter that works within forming requirements in order to control surface textures (Durante et al. 2010) (Fig. 2).

2.1 Dealing with Forming Inaccuracies

In SPIF, the most inaccurate regions are usually at the 'outline' of the formed part. The metal tends to elastically deform and move the geometry along the forming axis. This occurrence is especially apparent when forming parts that meet the flat stock at sharp angles. While it is possible retain the border of the desired geometry with a backing plate or partial die, this research has been mainly aimed at investigating the potentials of a completely die-less process.

To work with the issues that arise from SPIF additional tool paths and geometric modifications were tested. Initially, the same tool path was run multiple times on the same part; a strategy that has been proven to increase forming accuracy and help to reduce the maximum deviations up to −0.5 mm (Meier et al. 2009). This increased the accuracy of the part everywhere except for the perimeter, because this region was drawn past the desired depth. Modifications to the position of the part proved to increase accuracy of the edges. Adding geometric 'skirts' to the parts' edges stiffens the geometry and provides a deeper forming condition, hence reducing the overall deviation at the edge (Kreimeier et al. 2011).

Additional tool paths were tested other than the spirals. A zig-zag tool path method not only reveals where the actual part is on the panel but also draws the metal deeper and more consistently. The identifiable outline could be used to easily register the geometry for a secondary process such as trimming and panel registration (Fig. 3). When conventional secondary passes (spirals) were used, the outer portion of the geometry is formed while the center is moved inward leaving it touched with much less force by the forming tool. Since the zig-zag moves across the formed geometry linearly instead of radially, the center of the formed is

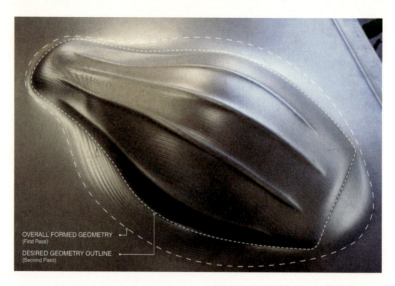

OVERALL FORMED GEOMETRY
(First Pass)

DESIRED GEOMETRY OUTLINE
(Second Pass)

Fig. 3 A refined part with overlapping spiral and linear toolpaths

not isolated by newly formed material, thus allowing the center to maintain its position relative to the forming tool which allows the forming tool to maintain contact with the material. The contrasting grain also provides a unique aesthetic pattern that could be exploited as part of the component design (Fig. 3).

2.2 Laser Scanning

While precise articulation of the formed parts can be achieved by running the same toolpath multiple times, their accuracy could only be determined by a digital scanning method. With the addition of a scanned digital representation of the formed part, the increase in accuracy of each pass could help validate forming results and detect slight variations in the geometry.

For the scanning routine, the robot is programmed to move to fixed points along a tool path and shoot a laser point at each position. When the tool is at the approximate center of its range, a voltage reading from -1 to -10 V is then converted to a range coordinate in millimeters and an XYZ position. All the information gathered from this process is stored in a separate file. The data files are then used as inputs in a custom script that redraws the toolpath to displays any deviations between the forming and scanning toolpath. It also shows where the formed parts are in relation to the digital model (Fig. 4). The scanning results for a test panel formed by SPIF (shown below) reveal a minimum deviation of 0.008 mm at the center of the part and as maximum of 13.75 mm at the edges.

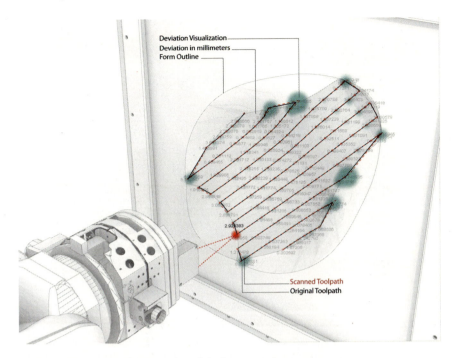

Fig. 4 A diagrammatic demonstration of the laser scanning process

This system is still currently being tested for repeatability and the effect of different surface angles on the laser sensor readings.

The model produced by the scanning process is then used for additional processes after the primary forming. It accurately locates where detailed passes should be positioned. The model is also measured at the edge of the designed part to verify and reconstruct a tool path for cutting the material from the stock sheet.

3 Dual Side Incremental Forming

Moving on from SPIF forming to DSIF presents a set of new challenges that couldn't be succumbed without the invaluable knowledge acquired from SPIF. Forming panels that have positive and negative Gaussian curvatures or rapid slope variations is very difficult using SPIF, however with DSIF there's more control and freedom over geometries that can be formed. Two different types of DSIF were employed and developed as the research moved forward (Fig. 5).

Fig. 5 *Top* DSIF Method A with a forming tool and a support tool. *Bottom* DSIF Method B with two forming tools

3.1 DSIF Method A: Forming Tool + Support Tool

This method requires programming a tool to continuously follow the perimeter of the form on the back side of the stock to stop the material from globally deflecting and moving the part along the forming axis. Constantly articulating the perimeter reduces forming inaccuracies at the edge even when there is a steep slope. This method eliminates the need for a backing plate with a cut-out to retain the 'flatness' of the stock at the formed parts' edges. A machined flat Delrin rod is used to fabricate this tool. Moving smoothly across the back side of the metal with low friction, the thermoplastic cylinder can resist forming pressures and prevent geometric creep (Fig. 5).

3.2 DSIF Method B: Forming Tool + Another Forming Tool

In order to support the material at each point in the forming process, a forming tool may be used from both sides of the material at the same contact point. When the forming tools are compressing the material at the points of contact, the accuracy and stability of the formed part are greatly increased (Malhotra et al. 2011). Laser scans of the perimeter show an averaging of the amount of deviation along the whole perimeter in DSIF whereas the deviation in SPIF shows significant fluctuation depending on the geometry as indicated in Fig. 6. Currently, the contact of both forming tools occurs during the beginning of the spiral. The gap between the two forming tools is determined by the sine law which has proven to inaccurately describe the material thickness causing the outer forming tool to deviate from the surface and eventually lose contact (Malhotra et al. 2011).

With SPIF, the generation of the tool paths is simple. The center points of the tool can be found by offsetting the designed surface by the radius of the tool and generating the forming spiral. With two tools, one cannot simply offset the surface in both directions, because the points along the spiral would not have been produced in a way that shares a contact point.

Toolpaths for this DSIF method are produced by calculating two separate tool center points that share a point of contact on the material. The tools' series of contact points must first be constructed to then find the positive and negative normal vectors from each point along the surface. Calculating the coordinated center points of the forming tools ensures that the two spherical tools stay tangent to each other along specific vectors (Maidagan et al. 2007) (Fig. 7).

The vector between the two forming tools can be modified to direct forces at the contact point if the material thickness is accurately calculated. The normal vector is the maximum angle off of the primary direction of the draw when forming. This process of constructing positions for DSIF is commonly used as it is a reliable method. The design of the spiral, through the use of Bezier graphed divisions,

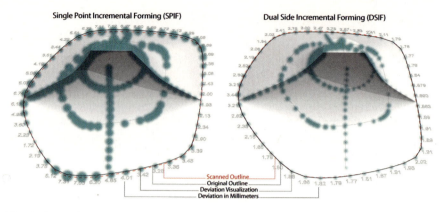

Fig. 6 Comparison showing edge and overall deviation between SPIF and DSIF

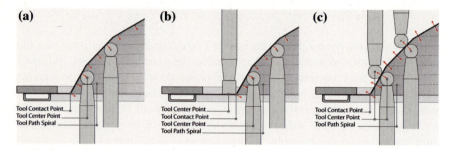

Fig. 7 a SPIF. **b** DSIF Method A. **c** DSIF Method B

allows for the forming process to be customized as an aesthetic and functional calibration of the spirals' density throughout the tool paths.

Pairs of points for each move of the robots are synced at every position through synchronous programming using KUKA Roboteam. Producing tools paths for more complex surfaces with both convexities and concavities has not been fully developed at this point in time due to the multiplicity of new factors that are introduced with DSIF. In the future, the ability to do so will open up the possibilities to produce more varied and topologically different panels.

4 Prototypes and Applications

Several proof of concept prototypes were produced as part of this research to demonstrate potential application of incrementally formed architectural skin components. A self-structured thickened porous skin was developed by forming multiple varied bell shaped components that change in size according to the

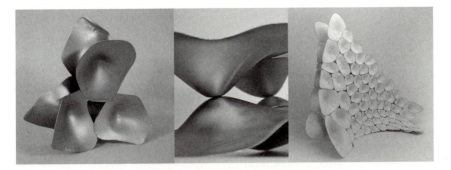

Fig. 8 Aggregation of self-supporting thickened porous skin

overall geometry curvature. The parts were spot welded together at the valleys and peaks to create the self-supporting system (Fig. 8).

Another system was developed as a series of panels that have a global pattern of ribs which structure the edge and register the panels through overlapping connections. The central protrusions are formed as undercuts (easily achievable with multi-axis forming) and orient the panel on a clipping system (Fig. 9). The bespoke ribs, bumps, and surface textures aren't formed solely on their aesthetic value, but are born out of the conflation of design, fabrication process, and connection detailing. In addition, they deliver a performative relationship between the panels and the materials formed (Hensel and Menges 2009). The aim is to avoid any kind of superficial patterning and allow for the expression of the performative aspects embedded in the panels. These patterns were a result of trying to find a design solution to minimize springback after trimming and utilized as indexing guides. A series of different structural patterns based on surface curvature analysis were developed and tested. Laser scanned models of the 'ribbed' trimmed panels show a 75 % improvement in the overall geometry when compared with untreated panels. The amount of deviation drops from 73 mm on the original un-textured panel to 15 mm after introducing the structuring textures (Fig. 10).

Another area of added functionality was geared towards developing panels that collect and direct water. The Namib Desert Darkling Beetle has the unique ability to 'fog-bask', a strategy utilized by the beetle to use its own body as fog collectors (Nørgaard and Marie 2010). Similar to the Namib Desert Darkling Beetles' elytra, the panels are designed to have areas coated with a hydrophilic, water attracting, film and other areas with a hydrophobic, water repellent, film.

The implication is creating a building facade or roof systems that employ such a technique to efficiently collect water from fog. For this strategy, 'micro bumps' are added to the formed component to increase the surface area. The hydrophilic peaks of the bumps allow the water to gather and then drip across the hydrophobic once it reaches its capacity. The surface texture not only structures the panel locally, but also become guiding flow-lines for shedding water in a controlled way. This system has been tested on small samples and is yet to be developed into a larger aggregated system (Fig. 11).

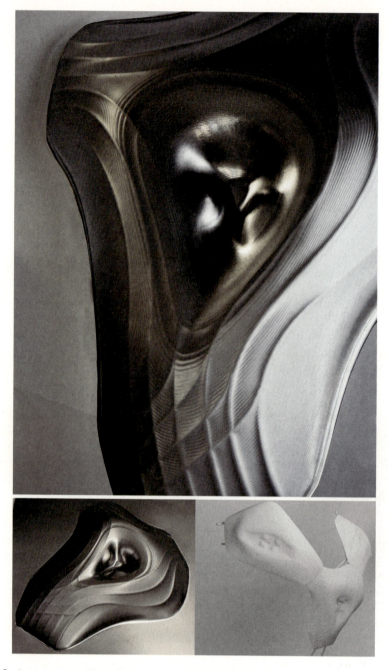

Fig. 9 A component with performative textures and features

Fig. 10 Comparison of overall geometric improvements with the 'ribbing' system

Fig. 11 Prototype of a textured performative water shedding surface

5 Conclusion

Most of the objectives in this phase of the research have been completed. Those include refining tool path generation, finding ways to incorporate forming inaccuracies in the design process, as well as developing a scanning and validation process to assist in forming secondary design and performative features. Developing a more robust double-sided forming process hasn't been fully achieved and is currently under development.

Given its low-cost and versatility, incremental sheet metal forming promises to radicalize the ways in which architectural sheet metal panels are designed, prototyped and fabricated. While this forming method's relative speed and efficiency is effective in the prototyping phase or for low-volume production, it still cannot compete with the high production speeds and precision of other contemporary metal forming processes.

Architectural precedents mentioned earlier serve an important starting point for developing such panels and enable an avenue to scrutinize existing paneling systems and study ways in which they could be advanced with a fabrication process like incremental sheet metal forming. The skins developed as part of this research were means to refine the process and its techniques rather than design fully fledged systems. Future research will utilize the knowledge gained from these studies to further advance and refine the process as well as investigate new possible types of performative aggregated cladding elements in a building envelope.

In conclusion, incremental sheet metal forming, in all its current variations, provides a unique opportunity to investigate new forms of architectural expression through performative varied façade systems. This research refines a closed loop method for an integrated computational design and fabrication process for incremental sheet metal forming as a means to produce low-cost parametrically customized aggregations of architectural skins.

Acknowledgments The authors would like to thank Taubman College FabLab Director Wes McGee for his continuous and generous support. This research was supported by the Taubman College of Architecture and Urban Planning at the University of Michigan, in addition to the Rackham Graduate School at the University of Michigan.

References

Durante M, Formisano A, Langella A (2010) Comparison between analytical and experimental roughness values of components created by incremental forming. J Mater Process Technol 210(14):1934–1941

Eekhout M (2008) Material experiments in design and build. In: Kolarevic B, Klinger K (eds) Manufacturing material effects: rethinking design and making in architecture. Routledge, New York, p 236

Hensel M, Menges A (2009) Patterns in performance-oriented design. Architectural Des 79(6):89–93

Ingarao G, Ambrogio G, Gagliardi F, Lorenzo RD (2012) A sustainability point of view on sheet metal forming operations: material wasting and energy consumption in incremental forming and stamping processes. J Clean Prod 29–30:255–268

Kalo A, Newsum MJ, McGee W (2014) Performing: exploring incremental sheet metal forming methods for generating low-cost, highly customized components. In: Fabricate: Negotiating design and making. gta Verlag, Zurich, pp 166–173

Kreimeier D, Buff B, Magnus C, Smukala V, Zhu J (2011) Robot-based incremental sheet metal forming—increasing the geometrical accuracy of complex parts. Key Eng Mater 473:853–860

Maidagan E, Zettler J, Bambach M, Rodríguez P, Hirt G (2007) A new incremental sheet forming process based on a flexible supporting die system. Key Eng Mater 344:607–614

Malhotra R, Cao J, Ren F, Kiridena V, Xia C, Reddy N (2011) Improvement of geometric accuracy in incremental forming by using a squeezing toolpath strategy with two forming tools. In: Proceedings of the ASME international manufacturing science and engineering conference, Dearborn, 30 Aug 2013, pp 603–611

Meier H, Buff B, Smukala V (2009) Robot-based incremental sheet metal forming—increasing the part accuracy in an automated, industrial forming cell. Key Eng Mater 410–411:159–166

Nørgaard T, Marie D (2010) Fog-basking behaviour and water collection efficiency in Namib Desert Darkling beetles. Front Zool 7(23):1–8. Available from: Frontiers in Zoology (30 May 2013)

Simmons M (2008) Material collaborations. In: Kolarevic B, Klinger K (eds) Manufacturing material effects: rethinking design and making in architecture. Routledge, New York, p 262

An Approach to Automated Construction Using Adaptive Programing

Khaled Elashry and Ruairi Glynn

Abstract With advancements in sensing and adaptive behavior programming the use of robots in complex environments such as building sites (typically the sole domain of human labor) is becoming increasingly feasible. However, large challenges remain before we can consider on-site robotic construction as a cost effective alternative to other means of construction. One major challenge lays in developing robotic systems that can safely work with unpredictable construction materials alongside human co-workers. To achieve this continuous sensing/actuation feedback is required, an approach that contrasts that typically found on an assembly line where "blind" robots perform pre-programmed routines. Our research presents a new robot control plug-in called "Scorpion" and a Java/Processing server, which in combination provide a workflow where multiple layers of sensing provide fast, continuous feedback to a robotic workcell. As a demonstration, this adaptive programming approach is employed in a case study of robotic bricklaying with mortar, using traditional tools as end effectors.

Keywords Industrial robots · Construction · Robotic control · Computer vision · Adaptive programing

1 Introduction

While robots operate with great speed and precision in controlled environments such as factory assembly lines or research laboratory, in the complex world of the construction site significant challenges remain. Tasks which are seemingly simple

K. Elashry (✉) · R. Glynn
Bartlett School of Graduate Studies, London, UK
e-mail: Khaled.elashry.12@ucl.ac.uk

R. Glynn
e-mail: r@ruairiglynn.co.uk

W. McGee and M. Ponce de Leon (eds.), *Robotic Fabrication in Architecture, Art and Design 2014*, DOI: 10.1007/978-3-319-04663-1_4,
© Springer International Publishing Switzerland 2014

for the human work force to achieve, like avoiding obstacles or handling hetero-geneous materials, are prohibitively difficult for robots, preventing wider adoption of these manufacturing technologies within the building industry. While the use of industrial robots has generated many extraordinary architectural design and research projects within laboratory environments, an ongoing question remains whether these technologies can be translated into on-site applications. One view is that robotics should remain an off-site process where components are produced and then transported to site; however, the possibility of on-site application remains tempting for many and requires further investigation.

In comparison to the construction industry, productivity in manufacturing industries that incorporate robots has increased exponentially in the last decade (Balaguer and Abderrahim 2008). Industrial robots were never intended to work in unpredictable environments, developed to do specific tasks with specific programs that are run for the lifetime of the machine (Keating and Oxman 2013). On-site industrial robots open the door for automated fabrication processes using materials available in proximity to the construction site, eliminating the burden of trans-portation while encouraging local construction.

At the end of the last century several attempts were made to integrate auto-mated robotics in construction. During the Japanese booming construction period in the 1980s up to 200 systems of automation were developed (Bechthold 2010), often large custom machines that requiring substantial expenses to build and operate. Research in automated construction continued during the 1990s, and in 1994 the 11th ISARC conference published several papers discussing automated construction. A system under the name of ROCCO was proposed as a solution for robotic masonry work on-site (Andres et al. 1994), referred to as a "Fault Tolerant Assembly Tool". As one may infer from the title it relied on sensor feedback for fault correction. Research in the platform continued with two additional papers in 1995 and 1996, but was never put in practical use. Recently several on-site sensor-informed systems have been tested like "DimRob" (Helm et al. 2012), a mobile fabrication robot under development at ETH, Zurich. This system informs the common industrial robot with sensor data, allowing it to perform mobile construction.

Machines negotiating complex environments are built upon decades of applied mobile robotics research and offer models that could be applied to design and fabrication. Sensing and feedback are key features of these systems, suggesting that any on-site application of robotics would likely require similar feedback mechanisms to cope with the uncertainty of these environments. Conventional workflows between design and digital fabrication involve a linear transfer from digital model to machine instructions. These instructions are typically executed "blind", lacking any feedback system. The rapid development in 2D and 3D vision systems offers interesting opportunities to build workflows that allow robots to "see" and adapt their behavior in real-time to changes in their environment. This type of feedback also allows for fabrication processes to incorporate heterogeneous components and materials with unpredictable behavior that would otherwise be difficult for robots to work with.

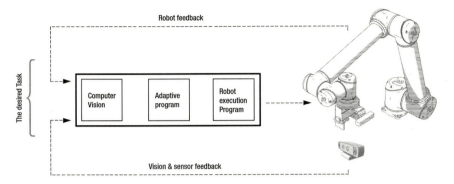

Fig. 1 Work flow system diagram

This chapter introduces a workflow (Fig. 1) that allows industrial robots to work in complex environments and deal with relatively unpredictable materials. This allows the robots to work on composite building jobs that humans require extensive training and practical experience to properly complete. In these conditions success is determined by the robot's ability to automatically shift from working with one material to the other and alter its program according to feedback data. The workflow we describe aims at simplifying the process of designing and testing such tasks through robot task planning, computer vision, and adaptive programing.

2 Industrial Robotic Arms from Universal Robots

The industrial robots used for the research were Universal Robots (UR), a range of small to mid-scale machines manufactured in Denmark by Universal Robots A/S. These novel industrial robots have several features that differentiate them from industrial robots from other manufacturers, such as the ability to operate via direct upload and safety features that allow them to work along humans. Additionally, the UR are built with an uncommon servo configuration referred to as a 'non-spherical wrist configuration'. While the spherical axis configuration found on most industrial robots has the three orientation axes intersecting at a single point, the UR are built with a shifted, non-spherical wrist configuration (Fig. 2). The core difference between the two configurations can be found in their respective non-linear inverse kinematics solvers. While the IK solver for spherical-wrist robots have a notably simpler solution, it can suffer from frequent singularities; the UR's IK solver has eight solutions, meaning any point in the UR's operational space can be achieved by nine different axis configurations (Fig. 2).

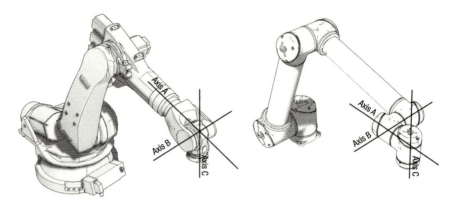

Fig. 2 Difference between spherical and non-spherical wrist robots

3 Robot Control Programing

The versatility of industrial robots are ultimately defined by their control software and the task they are programed to undertake. The robot needs an execution program written in a manufacturer-specific language which describes a procedural set. In an industrial setting these programs are typically produced using a manufacturer-supplied software. With the recent introduction of robotics in architecture, many robotic programing tools have been developed in response to the architectural field's need for a fast and easy way to control robots in familiar design environments. These controllers, like HAL (http://hal.thibaultschwartz.com/) and KUKA PRC (http://www.robotsinarchitecture.org/kuka-prc), presented a means to easily program industrial robots where the user generates a movement path with common CAD software and the programming tool generates the robotic execution program. Both these tools were built on the top of Grasshopper (http://www.grasshopper3d.com/), a Visual Basic.NET plug-in for the 3D NURBS modelling software Rhinoceros (http://www.rhino3d.com/).

At the start of the research there was no control software publically available for the URS, and existing tools built for other robots could not be used due to the difference in servo configuration. The absence of control software meant a one would need to be built from scratch. Building such a tool proved to be a complex task due to the lack of data available regarding Controlling the URs—The available controlling plug-ins also lacked the ability to improve upon it through the development of additional components or editing the existing ones due to thier proprietary nature. On the other hand software like Lobster by Daniel Piker (http://www.grasshopper3d.com/group/lobster) has taken a different approach, opening a door to open-source robot control in Grasshopper. Similar initiatives are gaining momentum in robotics-related fields with projects like ROS, the robot-operating program by Willow Garage (http://www.ros.org/). Unfortunately these projects still require the user to have a substantial knowledge of programing in order to use

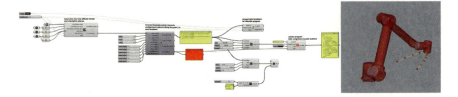

Fig. 3 Scorpion 0.1 grasshopper components and representation in Rhino viewport

them properly. With our Grasshopper component Scorpion (Fig. 3), we aspired to build a platform that combines the availability of predefined tools for new users with access to core functions for more experienced users. Its development focused on the following features:

1. open-source and free, allowing easy access for advanced users to easily adapt the software to their specific needs and contribute new tools and functions to the software
2. the ability for beginners to easily control the robot while still providing access to feedback tools
3. allows for direct control of the robot from the PC without the need to transfer files
4. allows for feedback from the robots to be read directly in the software.

In order to achieve these goals, Scorpion was developed to have multiple components, including: an IK solver, the robot upload component, the feedback component, and additional 'helper' components.

3.1 IK Solver Component

The input for Scorpion's inverse kinematic solver is the plane located at the tip point. Planes are objects created in Grasshopper which contain all the required data (three translations and the rotation) needed to calculate the kinematic chain of the UR. Starting with this input the kinematic chain is computed, then the joint angles are calculated accordingly. The inverse kinematic chain is calculated by solving for the location of three key points along the chain. Since each of these points has multiple solutions for a single input, three toggles were added to the input of Scorpion's inverse kinematic solver (Fig. 4) to switch between the nine different IK solutions (Fig. 5). Each servo in the UR operates within an angle range from $-360°$ to $+360°$, resulting in two legitimate solutions for each angle. This raises the number of solution for each input to 512 using six more angle toggles. These solution toggles allow the controller to shift from one solution to the other to avoid problems like collision, singularities, or attempting to move beyond the operable servo range.

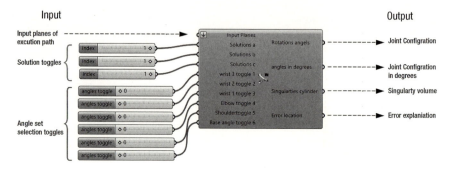

Fig. 4 The IK solver component in Scorpion

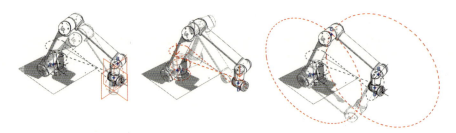

Fig. 5 The nine IK solutions for the universal robots

3.2 Robot Uploader Component

After the joint locations that represent the path are calculated, the "Move" component is responsible for converting these values to the URscript language, a set of instructions the Universal Robots can execute (http://www.wmv-robotics.de/home_htm_files/scriptmanual_en_1.5.pdf). The component also provides the user the option of changing between linear movements in either tool space or joint space. The movement commands are then combined with input/output states in the uploading component, which connects to the UR via TCP/IP and instantly executes the combined code (Fig. 6).

3.3 Feedback Component

The feedback component (Fig. 7) is considered an important aspect of Scorpion that differentiate it from other robot control plug-ins. The code for the feedback component is based on the Python library for UR control by Oroulet Roulet (https://github.com/oroulet/python-urx). By default Universal Robots communicate their current state back to the controlling PC via TCP/IP. The component allows for real-time visualization of the robot configuration from within the Rhino/

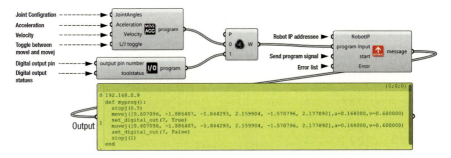

Fig. 6 Movement and upload components in Scorpion

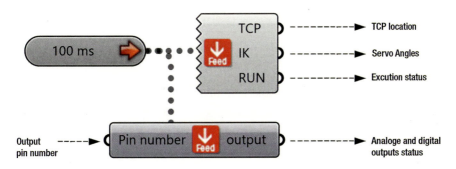

Fig. 7 Feedback components in Scorpion

Grasshopper environments by reading the states of the controller's analog and digital signal statuses, the location of robot's tool-tip; and the state of the program in execution. This component has proved useful when working with dependent consecutive programs that require the robot's start position be determined by its current location. The use of this component still has many additional possibilities that we intent to explore in future research initiatives.

4 Feedback Server and Automated Workflow

Scorpion was developed to be an independent, fully-featured robot control tool incorporating real-time feedback and visualization path recording and program-execution monitoring. However, problems were encountered with both computational speed and the difficulties of work with computer vision in Grasshopper. To achieve a reliable workflow, some of the automated control task were handled by a dedicated Java/Processing server (Fig. 8) which was responsible for:

- executing computer-vision routines on video and point-cloud data
- generating inputs for Scorpion's parametric programs according to sensor and computer-vision feedback data.

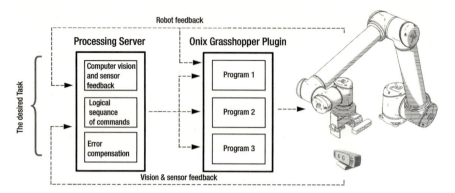

Fig. 8 Detailed workflow diagram

When a task is initiated the server builds the sequence of operations necessary for the task to be achieved using parametric robot programs designed in Scorpion. The server sends UDP protocol messages containing the input data for each of the parametric programs to Grasshopper one-by-one. The server waits for each program to execute successfully before sending the next data set, compensating for errors detected through computer vision and sensor feedback.

5 Bricklaying Automation

To test the workflow's ability to respond to a range of sensing and real-time decision-making processes, an on-site construction process involving unpredictable material behavior was chosen: bricklaying. While pick-and-place operations for brick wall assembly has been well documented by Gramazio and Kohler, their research replaced mortar with a bonding glue. Additionally, many of their masonry projects were produced in a precise lab environment, such as the Gantenbein vineyard façade project (Gramazio and Kohler 2008). More recent Gramazio and Kohler projects have explored the application on-site mortarless wall assembly as partitions or ornamental object.

Unlike robotically-built brick work that depends on pick-and-place operations and precise bonding material placement, our research aimed to build a brick wall using the same mortar and tools used by a human bricklayer. This requires a single robot to automatically shift between working with mortar (a relatively unpredictable material when in its liquid state) and brick placement. An end effector was designed around an industrial pneumatic gripper, allowing the robot to shift between the tasks of brick placement, 3d scanning, and mortar laying (Fig. 9). Two aspects of bricklaying were chosen as the main evaluation criteria of the test:

- the technique needed to properly lay mortar and place bricks
- an awareness of a larger scheme for error monitoring and movement compensation.

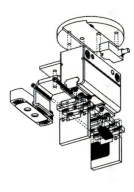

Fig. 9 End-effector design

The bricklaying process was divided into small task, each with parametrically constructed movement paths which update according to a variety of inputs. Some of these inputs are determined during the initial calibration phase while other change each time a task is executed. The programed tasks included:

1. pick and place of bricks
2. switching end-effector
3. using sensors and 3d depth camera
4. laying mortar
5. removing excess mortar after laying the new layer of bricks (Fig. 10).

Laying mortar was an interesting challenge within the process. After the pneumatic gripper picks up the masonry trowel, the end-effector is adjusted in Grasshopper, shifting the working plane from the middle of the gripper to the middle of the trowel surface. The success of the mortar-laying moment was determent by two factors: the ability to move all the mortar from the trowel to the brick surface, and the creation of a consistent layer of mortar on top of the brick.

A professional bricklayer usually uses one continuous movement to lay the mortar on a brick. After observing bricklayers' gestures, it was noted that there are numerous factors which adjust the movements of the trowel, such as the amount of mortar on the trowel and the consistency and weight of mortar. Since mortar consistency changes rapidly as its water content changes, the bricklayer unconsciously adjusts his movements according to his training.

The UR's teaching mode allowed the robot to be physically manipulated while streaming joint positions through the TCP/IP connection; these movements were then recorded in real time through Grasshopper to create movement paths. This allowed a tactile method for documenting the basic gestures of a bricklayer, which could later be used by the robot. The stored movements were then iteratively develop to develop a global success rate for mortar laying. The input for the task is

Fig. 10 The robot removing excess material after laying the bricks

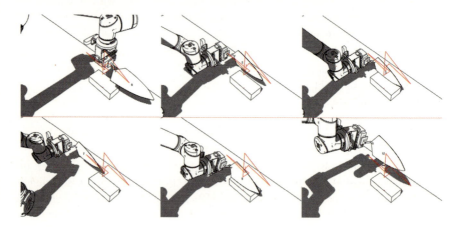

Fig. 11 Mortar laying movements

the location of the mortar (determent by computer vision) and the location and rotation of the target brick (Fig. 11).

5.1 Tasks of Computer Vision and Sensing

Computer vision was an important part of the process due to the material unpredictability of the mortar caused by its shift from a viscous liquid to solid. The unpredictability made it challenging to get the desired amount of mortar on the trowel and onto the brick, a task that required a means to determine the arbitrary surface of the mortar before attempting to manipulate it with the trowel (Fig. 12).

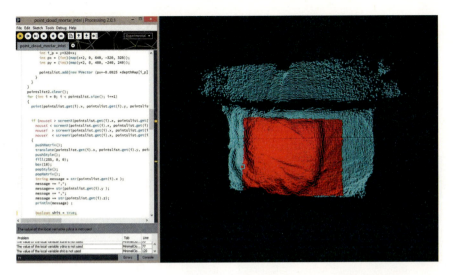

Fig. 12 Depth camera point cloud data calibration

Depth cameras provided a possible solution to this problem. A vision systems overlooking the process could potentially have 'blind spots' caused by obstructions in its line of sight. To avoid this problem, a camera with depth-sensing capabilities would need to be mounted directly to the end effector. Intel's new Perceptual Computing camera was used due to its relatively small form-factor. Unlike the more common Microsoft Kinect, the Perceptual Computing camera works within a range of 0–1.5 m, making it possible for the robot to move around and 'peek' at objects to obtain a high-density point cloud of their geometry. A surface-approximation algorithm was used to calculate the volume of the mortar, approximating each point as a volumetric box. This was used to determine the ideal trowel depth required to extract the necessary amount of mortar, which was then fed back into the robot program that handled moving the mortar.

Additional sensing routines were incorporated into the workflow to check the height and levelness of placed bricks. Initially the depth camera was used to check the height of the brick after placing it on the mortar surface, but this was deemed unreliable as the noise in the point cloud data prevented measurements of less than 1 mm. In its place a limit switch connected to the digital input of the robot controller was added to the end effector. During operation the robot would move towards the surface of the brick until the limit switch was pressed, recording the location of the new point in the server program and updating the virtual model with the new data. This approach was able to accurately determine any variation between the computationally-derived brick position and the actual position of the corresponding placed brick. This information was crucial in determining the error factor needed to build a sample of a large wall. However the process suffered from speed issues as it was necessary for the robot to test many individual points, adding to the run time for the task.

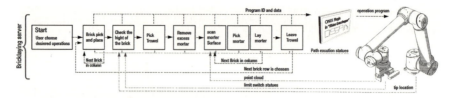

Fig. 13 Operational sequence of the bricklaying process

Fig. 14 First wall sample test

5.2 Final Tests of the Brick Laying Server

Using a set of parametric programs created in Scorpion that describe each task in the process, from picking up a brick to 3D scanning, the server builds a logical sequence of tasks and executes them sequentially, integrating data from sensors and feedback into the workflow (Fig. 13).

The outcome of this initial test of the workflow's ability to incorporate a range of sensing and real-time decision making processes enabled the robot to autonomously build a series of sample walls. The first prototype was 29 cm high and 111 cm wide, consisting of 4 courses of full and half bricks (Fig. 14). The programs were generated to work according to a location relative to the robot, allowing the program to be easily executed in any location without the need to alter the program. The height of each course was detected with the end effector's limit switch, allowing the server to automatically compensate for height errors before laying a new layer of mortar.

A second test was conducted to test the program's ability to build different design formations (Fig. 15), with the rotational position of each brick changed to create a pattern which generated both the pick-and-place and mortar laying programs. As the server would need to alter the pick location of the new brick avoid collision between the adjacent bricks and the pneumatic gripper, an algorithm was added to determine the best location to pick up a brick according to its angle of placement. The final constructed sample was 22 cm high and 75 cm wide. Although the end results of the two tests were not perfectly finished, they showed

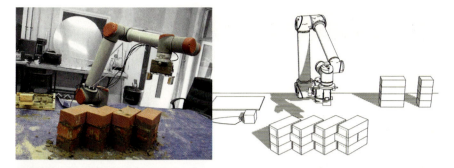

Fig. 15 Second wall sample test

that the a workflow incorporating Scorpion as a parametric planner and a server for computer vision and adaptive programing can be used to achieve composite tasks such as bricklaying.

6 Conclusion

This research presents a workflow that allows a robot to work on a task where material complexity might otherwise prove prohibitive. While still produced within the relatively controlled environment of a research lab (rather than the considerably more complex on-site environments it is intended to work within) the success of bricklaying coupled with 3D scanning and proximity sensing demonstrates a promising proof of concept. The use of mortar will eventually allow robots the ability to build structural walls on-site in addition to cladding systems. The research also demonstrates how sensing and real-time decision making allows robotics to perform processes typically associated with a human work force, the implications of which are beyond the scope of this chapter.

This chapter also presents Scorpion—an open-source Universal Robot control plug-in for Grasshopper. Its unique features including direct upload and feedback control, in addition to parametric path programing and increasing the accessibility of industrial robots. Although the development of the plug-in is still in its initial stages, this chapter presents its promising potential as a control platform. This chapter does, however, omit how Scorpion has proven to be a successful teaching tool for students at the faculty it's being developed at.

Both the end-effector and the robot itself offer data feeds not yet incorporated into the workflow, including live video feed from the Intel camera and torque data from the robot servos. This could offer an additional layer of information to drive decision making as further layers of sensing are introduced. Machine learning will become a logical next step towards more sophisticated real-time adaptive robotic behavior, producing a new generation of robotics not only adaptable to unpredictable material behavior, heterogeneous materials and complex environments, but potentially also adaptable in their fabrication strategies.

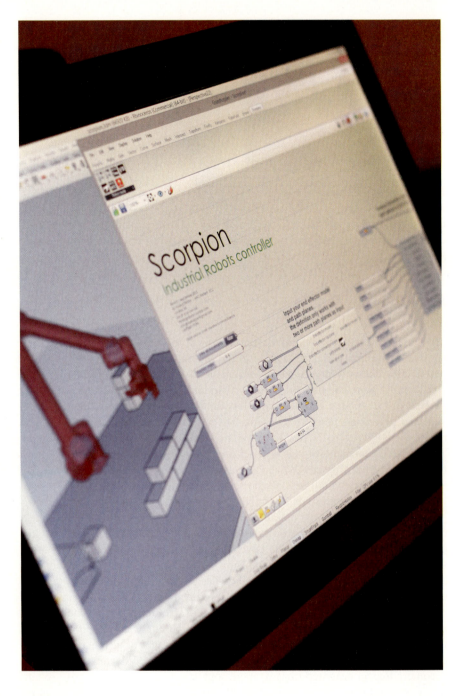

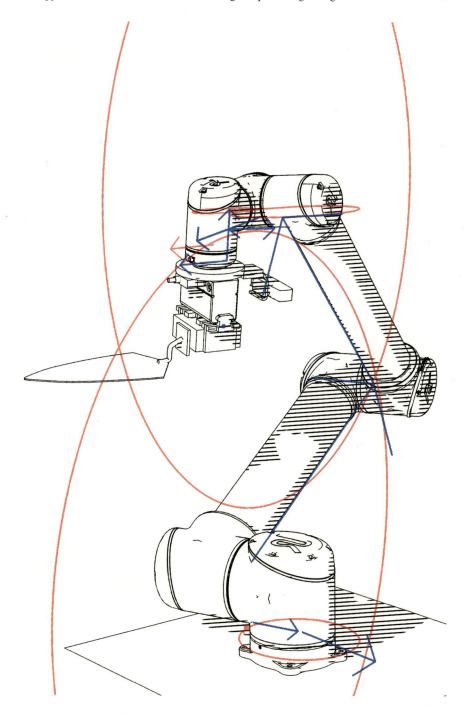

References

Andres J et al (1994) First results of the development of the masonry robot system ROCCO: a fault tolerant assembly tool. Autom Rob Constr XI 11:87–93

Balaguer C, Abderrahim M (2008) Trends in robotics and automation in construction. Rob Autom Constr 1–22

Bechthold M (2010) The return of the future: a second go at robotic construction. Architect Des 80(4):116–121

Gramazio F, Kohler M (2008) Digital materiality in architecture: Gramazio and Kohler. Lars Muller Publishers, Boden

Helm V et al (2012) Mobile robotic fabrication on construction sites: DimRob. In: 2012 IEEE/RSJ international conference on intelligent robots and systems. IEEE, pp 4335–4341

Keating S, Oxman N (2013) Compound fabrication: a multi-functional robotic platform for digital design and fabrication. Rob Comput-Integr Manuf 29(6):439–448

Design and Fabrication of Robot-Manufactured Joints for a Curved-Folded Thin-Shell Structure Made from CLT

Christopher Robeller, Seyed Sina Nabaei and Yves Weinand

Abstract The prototype presented in this chapter utilizes the technique of curved folding for the design of a thin-shell structure built from curved cross-laminated timber panels (CLT). The curved-folded geometry allows for a span of 13.5 m, at a shell thickness of only 77 mm. The construction requires curved line CLT joints, which are difficult to address with state-of-the-art jointing techniques for CLT. Inspired by traditional woodworking joinery, we have designed connections for the integrated attachment of curved CLT panels, utilizing digital geometry processing tools to combine the advantages of traditional joinery techniques with those of modern automation technology.

Keywords Curved folded plate structures · Robot fabrication · Cross-laminated timber · Mono-material

1 Motivation

Recent studies about folded plate structures such as the Chapel St. Loup (Buri and Weinand 2010) and the "ICD/ITKE Research Pavilion 2011" (Magna et al. 2013), have demonstrated architectural and structural applications of folded-plate structures built from plywood and cross-laminated timber panels (CLT). Architectural

C. Robeller (✉) · S. S. Nabaei · Y. Weinand
EPFL Laboratory for Timber Construction IBOIS, Lausanne, Switzerland
e-mail: christopher.robeller@epfl.ch

S. S. Nabaei
e-mail: sina.nabaei@epfl.ch

Y. Weinand
e-mail: yves.weinand@epfl.ch

W. McGee and M. Ponce de Leon (eds.), *Robotic Fabrication in Architecture, Art and Design 2014*, DOI: 10.1007/978-3-319-04663-1_5,
© Springer International Publishing Switzerland 2014

Fig. 1 Full-scale, mono-material prototype structure, spanning 13.5 m at a shell thickness of 77 mm, exhibited at the academy of architecture in Mendrisio, Switzerland

applications of curved-folded geometry have previously been studied for benefits such as the panelization of complex shapes (Kilian et al. 2008).

The curved-folded thin-shell structure presented in the present chapter builds up on this research and demonstrates a structural application of *curved CLT*, a relatively new but increasingly popular product offered by European CLT manufacturers. Our prototype (Fig. 1) utilizes the curved shape of the panels to increase the inertia of the structure in its main load-bearing direction, acting very much the same way as a bi-cantilever frame. Our interpretation of the traditional simple frame is meanwhile innovative using curved thin-shell panels instead of beams, which are transmitting structural efforts in a curved line of action through robot fabricated wood working joints.

Moreover, connecting two reversely curved panels through a curved fold would propagate the curvature from one into another when one of them is bent. This interaction helps resisting lateral forces which tend to flatten their curvature. These lateral forces are transformed into out of plane forces due to the shell behavior of panels, at the interface and transmitted through the curved panels until the clamped edge.

Alternatively, such deformations are supported in practice by tensile steel cables along the chord of the curved CLT elements, as demonstrated in the recent wide-span roof construction of a factory building in Ainet, Austria (http://www.holzbau-unterrainer.at/Hallenbau). Instead of tensile cables, our prototype supports these forces by taking advantage of the curved folding technique.

Unlike origami paper folded forms, wood panels cannot actually be "folded". Instead, the performance of such structures depends on linear edge-to-edge connections between the panels, making these joints a *key component* in the design of wooden folded-plate structures.

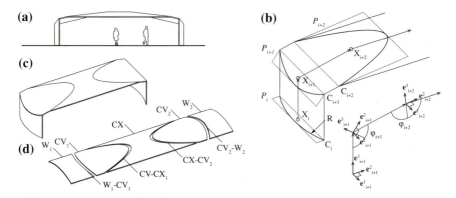

Fig. 2 **a** Full-scale prototype side view schematic. **b** Curved-folding technique. **c** Curved-folded thin-shell geometry. **d** Unfolded CLT parts/fabrication layout

2 Project Geometry

As illustrated in Fig. 2b, the curved-folded geometry is generated through a circular arc C_i (Radius R = 5.9 m) and its reflection about a set of planes P_i. These reflection planes are located at the nodes X_i of a polyline, which lies on a vertical plane normal to the chord of C_i. Each Reflection Plane P_i is normal to the bisector of the two neighbor segments of the polyline. The fold angles φ_i are determined by the interior angles between the polyline segments. We used this polyline as a variable control polygon in a parametric design model.

The *unfoldable* geometry (Fig. 2d) allows for an efficient fabrication without offcuts. Depending on the length of the panel manufacturer's production line, all parts can be produced from a single, 17 m long curved CLT panel.

3 Limitations of State-of-the-Art Connectors

Figure 2d illustrates the four connections W_1–CV_1, CV_1–CX, CX–CV_2 and CV_2–W_2 that are required for the assembly of the curved-folded CLT thin-shell. The curved panels must be jointed at precise, non-orthogonal angles, which are difficult to address with state of the art CLT jointing techniques: Glued-in metal connectors, which are common in glulam and LVL constructions, cannot be applied to CLT because of its internal stress cuts, into which the adhesive could discharge.

Alternatively, external metal plate connectors could be applied visibly on top of the panels with wood screws. A large amount of such connectors would be required in order to achieve a sufficiently rigid and stiff joint. In general, the before mentioned connectors reduce the stiffness of a folded-plate, while modern wood adhesives, such as 1K-PUR, allow for a lossless, fully stiff joint.

However, the gluing procedure requires a fast and precise alignment and clamping of the parts, within the open time of the adhesive. For joints on orthogonal edges and flat panels, standard tools can be used for the alignment and clamping. However, non-orthogonal edges and bent or curved panels will require custom made guides or supports.

4 Joint Types and Algorithms

Inspiration for an integrated solution for alignment and attachment can be found in traditional wood joinery techniques such as *dovetail joints*, on which the prismatic edge geometry results in a single-degree-of-freedom (SDOF) joint, which predetermines the alignment and orientation of the two connected parts. Due to their involving manual fabrication, these traditional joints have lost importance during the industrialization. Later on, early mass-production technology was incapable of fabricating such connections.

Modern robot- and CNC fabrication technology however allows for the reconsideration of these techniques. An application on plywood panels has been demonstrated recently (Simek and Sebera 2010) as well as a spotfacing technique for the robot fabrication of finger joints on plywood (Magna et al. 2013).

On our curved-folded prototype structure, we have to address two different types of folds. For the joints W_1–CV_1 and CV_2–W_2 each with a fold angle φ of 102.7°, we have chosen a SDOF dovetail-type joint geometry (Fig. 3).

The geometry is generated by an algorithm, which divides the joint edge into an odd number of segments, returning a set of points X_i. A local frame $\{\mathbf{u}_i^1, \mathbf{u}_i^2, \mathbf{u}_i^3\}$ is added to these points, where and $\mathbf{u}_i^2$ is parallel to the $\mathbf{e}^2$ and $\mathbf{u}_i^1$ is the direction of insertion for the joint. These frames are rotated about $\mathbf{u}_i^1$ at an alternating angle $\pm\theta$. For each frame, a plane P_{Xi}, normal to $\mathbf{u}_i^3$ is intersected with the curves L_1, L_2, L_3, L_4, which gives us the curve and line segments for the dovetail joint geometry (Fig. 3b).

For the fabrication of these joints, we have chosen a *side-cutting* technique, using a shank type cutter (Koch 1964). The cutting requires simultaneous 5x machining, for which we provide a set of tool center points TCP_i, each with a direction $\mathbf{m}_{orientation}$ (Fig. 4, right side). We have obtained these points and vectors through an algorithmic evaluation of a set of parametric surfaces S_i.

The algorithm detects inside corners automatically and adds notches (Fig. 4, right side), which allow for the insertion of the joint's counterpart and reduce the notch stress (Simek and Sebera 2010; EN 1995-1-1 2004). The transition of the cutter between joint faces with different orientations is solved through the bisector-planes P_i.

As one of its main advantages, the side-cutting technique allows for the use of simple supports under the workpiece. It is not necessary for the edge to cantilever. The main limitation for this method is set by the geometry of the cutter and the tool holder: The inclination β of the tool must be limited to avoid a collision

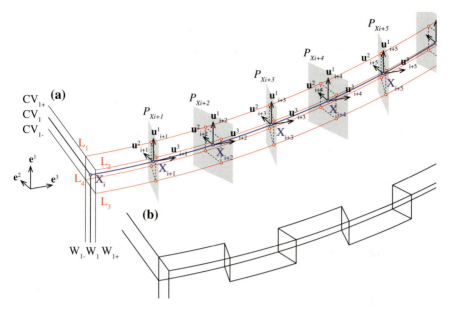

Fig. 3 Generation of the dovetail-type joint SDOF geometry through intersection of a set of planes P_{Xi} with the four curves L_1, L_2, L_3 and L_4. **a** CV1 is the part, the + and - show the intersection layers and W1 is the part, the + and - show the intersection layers **b** it is the same parts once jointed. Part CV1 above and part W1 below

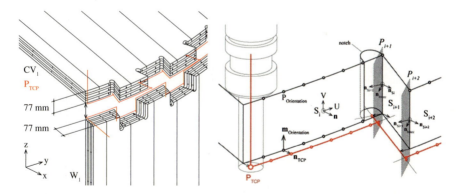

Fig. 4 Dovetail-type connections W_1–CV_1 and CV_2–W_2 on the full-scale prototype. *Left* schematic drawing showing the CLT panels and joints including layers and stress cuts, *Right* principle of machine code generation on a curved joint, using parametric surface evaluation

between the cutter and the wood panel. In consequence, this method can be applied to fold angles φ within the range of $90° \pm \beta$. With $\beta = 40°$ in our case, we are limited to folds up to $\varphi = 130°$. For the Connections CV_1–CX, and CX–CV_2, each with fold angle 167.2°, we have generated the joints with the method described in Fig. 3, but without a rotation about θ, giving us a planar joint geometry. In order to

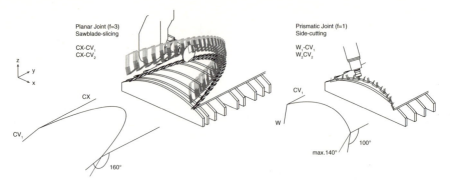

Fig. 5 Joint fabrication (tool path plot). *Left* planar "finger joint-type" connection (connections CV–CX), *Right* prismatic "dovetail-type" joint (connections W–CV)

cut the open slots of these joints, we used a saw blade-slicing technique instead (Fig. 5, left side), making multiple vertical cuts for each slot. This method allows us to fabricate very obtuse folds on simple, non-cantilevering supports.

5 Preliminary Structural Analysis

A preliminary structural analysis task has been performed in order to determine an initial value for the panel thickness and assess structural performance of the curved geometry to support relevant loads. Results from this primary engineering stage would let choose (or design) the appropriate laminate product as well as to ensure that enough degrees of freedom are linked in order to properly transmit structural effort between panels. A Finite Element shell model has been established from the mid-surface of panels, with local joints between curved panels simulated as rotational and translation links acting on local axis. For each joint the local axis **x** is the tangent to the joint 3D curve, the **y** axis is the bisector of two normal vectors of connecting mid-surfaces. The z axis is then defined by the right hand rule.

Panels at this preliminary structural design stage are considered as isotropic with Young Modulus E = 7,800 MPa (a conservative value near to long term modulus of timber) and Poisson ratio $v = 0.3$. Structure is a temporary installation loaded for combination of self-weight and lateral as well as transversal wind load cases. Results for Ultimate Limit State and Service Limit State propose 77 mm thickness to be a safe initial size for the prototype (Fig. 6). This preliminary dimension will be confirmed through a detailed model once connection technology and the panel laminate product have been selected through the investigations which would be carried out on the 1:5 scale prototype.

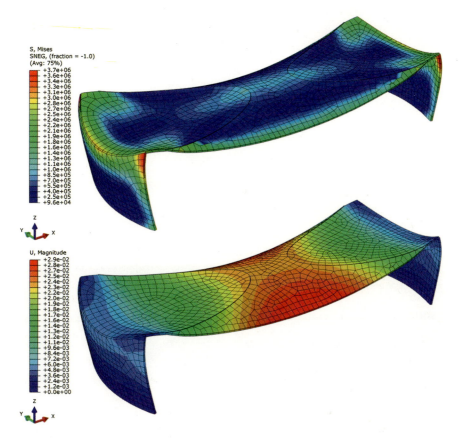

Fig. 6 Preliminary structural analysis results under the self-weight load case. *Top* Von-Mises stress distribution. *Bottom* total deformation

6 Prototype at Scale 1:5

The fabrication methods and the assembly of the parts were tested on a smaller prototype at scale 1:5 before applying it to the full-scale prototype. The curved CLT, which is not available at the scaled thickness, was replaced by a laminate of thin plywood panels. A MAKA mm7s 5-axis CNC router was used for the formatting of the panels and for the cutting of the joints.

Due to the curved geometry of the parts, custom-made supports (Fig. 7) were required for the curved panels, which served simultaneously as a mold for our curved laminate.

A convex support was built for the parts CX, W_1 and W_2, and a concave support for the parts CV_1 and CV_2. An additional concave support was built for the saw blade-slicing technique, which requires access from above on the convex and the concave panel. The molds were fabricated from laminated veneer lumber panels

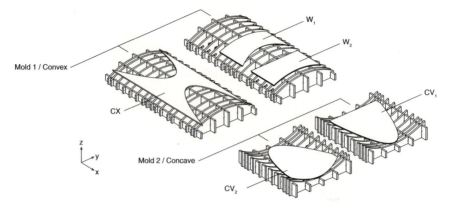

Fig. 7 Concave and convex mold for the lamination and formatting of the curved panels at scale 1:5

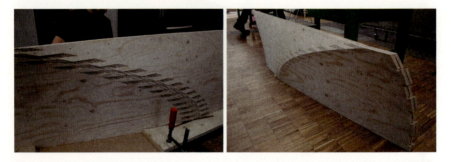

Fig. 8 Planar finger-type joints were used for the connections CV_1–CX and CV_2–CX. These joints were fabricated with a saw blade-slicing technique on a 5-axis CNC router

(LVL) and assembled with half-lap joints. In order to provide a precise calibration of the supports, a previously added oversize was removed through spot facing along the ribs on the CNC-router.

Figure 8 shows the assembly of the parts CX and CV_1 on the small-scale prototype: The short segments of the planar joint provide a large contact area and gluing surface between the panels.

7 Full-Scale Prototype

With the initial thickness determined at the preliminary engineering stage (see Sect. 5), a special laminate product has been designed by the manufacturer to fulfill the requirement regarding the thickness/radius ratio. The selected CLT panels are five layer symmetric laminates with total thickness of 77 mm.

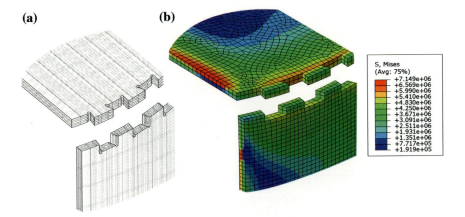

Fig. 9 Von-Mises stress distribution for Dovetail-type connections CV_1–W_1 and CV_2–W_2 on the full-scale prototype. Results are generated from a full three dimensional finite element simulation

As the next step, a full three-dimensional Finite Element model has been developed with this precise layer configuration and the mechanical properties provided by the manufacturer. The glued areas over the connectors have been considered as surface tie constraints. The goal of this detailed engineering step was to investigate local panel interaction over the connectors as well as to confirm the thickness of the CLT panel regarding the structural safety. The structural analysis confirmed the initial design thickness. The design issue was mainly to overcome deformation due to Service Limit State load combinations. The local and overall Von-Mises stress distribution is shown respectively in Figs. 9 and 10.

Having verified the structural parameters of the full scale prototype through the detailed analysis, the fabrication process was carried out at the facilities of the manufacturer. Figure 11 shows the fabrication of the full-scale prototype at the CLT panel production and processing facilities. The KUKA KR250 robot router with an additional linear-axis table allowed for the precise 5-axis fabrication of the joints at full scale. Its large workspace with a length of more than 30 m was essential for the realization of the prototype. The sides of the curved CLT panels were formatted with an 800 mm saw blade (Fig. 11, right side) and the dovetail-type joints W_1–CV_1 and CV_2–W_2 were cut with a 25 mm turn-blade knife shank-type cutter.

For the joints CV_1–CX, CX–CV_2, a different joinery technique was used for the full-scale prototype: 30 slots, each with a width of 30 mm were cut into the curved panels to be connected. Subsequently, connector elements with rounded edges (Fig. 12, right side), were fabricated from cross-laminated spruce LVL with a Hundegger K1 joinery machine. During the assembly of the full-scale prototype, these connectors were first used to align the roof segments, before gluing them in with a 1K-PUR adhesive. This alternative technique was chosen to reduce the machining time at full-scale. The robot router's higher sensitivity to vibrations required a lower feed rate compared to the CNC portal router (factor 0.2), while

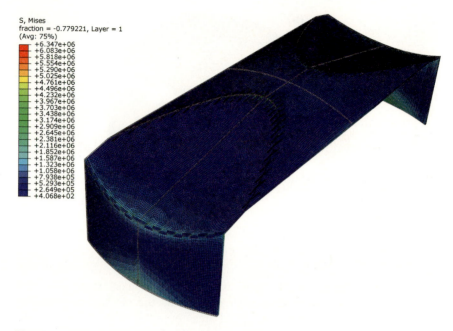

S, Mises
fraction = -0.779221, Layer = 1
(Avg: 75%)
+6.347e+06
+6.083e+06
+5.818e+06
+5.554e+06
+5.290e+06
+5.025e+06
+4.761e+06
+4.496e+06
+4.232e+06
+3.967e+06
+3.703e+06
+3.438e+06
+3.174e+06
+2.909e+06
+2.645e+06
+2.381e+06
+2.116e+06
+1.852e+06
+1.587e+06
+1.323e+06
+1.058e+06
+7.938e+05
+5.293e+05
+2.649e+05
+4.068e+02

Fig. 10 Von-Mises stress distribution for detailed model

Fig. 11 Robot fabrication of the full-scale prototype. *Left* fabrication of the joint CV$_1$–CX using a 25 mm shank-type cutter. *Right* formatting of the side of the panels using a saw blade with a diameter of 800 mm

the length of the tool paths increased by the factor 5 compared to the prototype at scale 1:5.

The slots for the LVL connectors could be fabricated without access from above on the concave panels CV$_1$ and CV$_2$. This allowed for all formatting and joinery to be done on one convex support. An additional concave support was not necessary for the fabrication of the full-scale prototype.

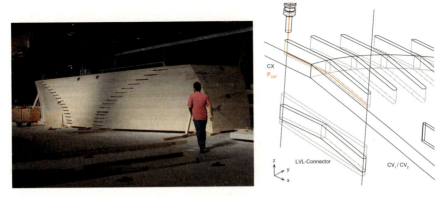

Fig. 12 *Left* assembly of the roof panels: shell prototype ready for the insertion of the LVL connectors (connections CV–CX). *Right* schematic drawing of the connector elements made from cross-laminated LVL panels. The slots have been cut with a 30 mm shank-type turn blade-knife cutter

Fig. 13 Assembly of the dovetail-type connectors between the wall panels and the roof panels (Connections CV–CX)

After the panel formatting, the different parts of the pavilion were assembled. The three parts of the roof CV_1, CX and CV_2 were assembled in a first step and lifted with a portal crane. Afterwards, the vertical elements W1 and W2 were positioned and an adhesive was applied to the faces of the dovetail-type connections W_1–CV and CV_2–W_2. Finally the complete roof was lowered into position from above with a portal crane (Fig. 13).

For weather protection, the entire shell has been treated with preservation for coniferous wood, before it was transported to its destination at the Academy of Architecture at the Universita della Svizzra Italiana (USI). After its final assembly on site, the structure was presented to the public in September 2013. To date, the temporary design is still exhibited on the campus of USI. Its exposure to the outdoor conditions is being documented.

8 Conclusions/Towards an Integration of Joints

Instead of using additional stiffening elements, the thin-shell presented in this chapter achieves its wide span with an integrated approach. The design utilizes the geometry of curved CLT panels and combines it with the technique of curved-folding.

The approach of integrated performance has been continued on the design of the joints: The connections were fabricated during the formatting of the panels and performed simultaneously as guides for the positioning and alignment of the pieces. The precise fabrication and the mono-material approach allowed for a visible application of the joints. This was not only useful for the aesthetic appearance of the prototype, but also for the simple monitoring of the condition of the joints.

Three different techniques were used for the joinery, all of which take advantage of computational fabrication technology for the fabrication of the joints and utilize digital geometry processing tools for the generation of both the joint geometry and the machining code.

The *side-cutting technique* used for the fabrication of the dovetail-type W–CV joints demonstrates an integrated jointing method for the connection of curved CLT panels. Compared to a butt joint or miter joint geometry, it offers advantages such as a precise and fast alignment and the machining time is only slightly increased. The technique allows for the jointing of fold angles within a range of 50° and 130°.

The *saw blade-slicing technique* allows for the jointing of more obtuse fold angles, but it requires additional access from above on the concave side of one panel. Both a convex and a concave support are required. The number of saw blade cuts for the open slots of the joints results from the total edge length divided by the saw blade thickness. Fast routers with a high feed rate are beneficial for the fabrication of this type of joint.

The *LVL connector technique* demonstrates an alternative method for the jointing of curved-folds with obtuse fold angles φ. It allows for fast fabrication and all joints were cut on a single convex support, but the manual assembly of the LVL connectors is more time-consuming compared to the dovetail joints.

While each of the presented joinery techniques shows certain advantages and disadvantages, we found the side-cutting technique to be a particularly effective and universal method for the jointing of curved CLT panels with a robot router. In general, the robot-manufactured joints allowed for the realization of our prototype, but we believe the construction demonstrates only one of many possible applications for curved-folded CLT structures.

References

Buri H, Weinand Y (2010) Origami aus Brettsperrholz, DETAIL Zeitschrift für Architektur + Baudetail 2010(10):1066–1068

EN 1995-1-1:2004 (2004) Eurocode 5: design of timber structures—part 1-1: general—common rules and rules for buildings, p 42

Kilian M et al (2008) Curved folding. ACM Trans Graph 27(3):1–9

Koch P (ed) (1964) Wood machining processes, collections: wood processing series. Ronald, New York

Magna R et al (2013) From nature to fabrication: biomimetic design principles for the production of complex spatial structures. Int J Space Struct 28(1):27–39

Simek M, Sebera V (2010) Traditional Furniture Joinery from the Point of view of Advanced Technologies. In: Proceedings of the international convention of society of wood science and technology and United Nations economic commission for Europe—Timber Committee, Geneva, Switzerland, 11–14 Oct 2010

Robotic Bead Rolling

Exploring Structural Capacities in Metal Sheet Forming

Jared Friedman, Ahmed Hosny and Amanda Lee

Abstract The robotic workflow proposed analyzes the bead rolling process, its potential digital interpretation, and improved fabrication aspects that accompany such a translation. For this process, a robotic tool has been developed that integrates multiple variables observed from existing bead rolling machines, while simultaneously allowing further control. Material-informed decisions required a series of tests evaluating optimum tool and workflow design. While the process provokes a multitude of potentialities, it has been put towards a structural behavior testing scenario to demonstrate its validity. It attempts to bridge analysis methods with prototyping as means of direct performance testing and evaluation. Deeply rooted within a parametric modeling environment, the workflow creates a single digital interface that links several platforms that otherwise are not in direct communication.

Keywords Bead rolling · Forming · Metal · Robot tooling · Mechanism

1 Introduction

The research conducted over the course of this project looked at taking what is typically a very manual process and transferring it to a robotic work environment to allow for high degrees of control. Bead rolling is a process of metal forming

J. Friedman (✉) · A. Hosny · A. Lee
Harvard Graduate School of Design, Cambridge, MA, USA
e-mail: friedman@gsd.harvard.edu

A. Hosny
e-mail: ahosny@gsd.harvard.edu

A. Lee
e-mail: alee1@gsd.harvard.edu

W. McGee and M. Ponce de Leon (eds.), *Robotic Fabrication in Architecture, Art and Design 2014*, DOI: 10.1007/978-3-319-04663-1_6,
© Springer International Publishing Switzerland 2014

consisting of two rotating dies that a person maneuvers a sheet of metal through in order to create grooves in the sheet. The application is most typically used in the automotive industry and for decorative embossing in metal plates. The specificity of the process requires a person with a high degree of skill to accurately move a sheet in and around the dies in order to achieve a desired bead in the sheet. By transferring the process to a robotic work environment, this research looks at the potential opportunities that are granted as a result of the control, speed, and replicability allowed by industrial robotic processes. A series of tooling prototypes were developed to explore these potentials and gain a better understanding of the opportunities granted by the process. Through prototyping with both manual and robotic processes, the authors recognized that there was a correlation between the beading patterns and the structural capacities of the sheets. This was identified as one of the primary opportunities to explore, and one which can serve broader applications when applied at an industrial level. The use of metal sheet forming techniques to provide strength to an assembly has been most commonly used in constructing curved shells or folded plate structures. While the bead rolling process may sometimes result in slight deformations of the of the sheet, the authors' intent here is to maintain the flatness of the sheet while investigating both the structural and ornamental opportunities enabled by advanced CAD/CAM processes. A digital workflow has been developed that automates a beading pattern based upon specific structural criteria such as location and magnitude of a load to be applied to a surface. The design-to-fabrication workflow is done entirely within the Rhinoceros modeling environment, using various Grasshopper® components specific towards structural analysis and robotic fabrication. As prototypes continue to be developed, empirical testing is being used to determine the levels of structural improvements that the process is capable of at the current scale of application.

2 Analysis of Existing Bead Rolling Processes

Existing bead rolling processes have remained essentially unaltered since their inception. Because of the forces required to push a sheet of metal through the dies, bead rollers are typically very rigid and made of cast iron or some other homogenous metal assembly. Industrial bead rollers can typically bead up to 14 gauge aluminum or 19 gauge stainless steel, whereas more consumer models will be able to handle 16 gauge aluminum or 21 gauge stainless steel (www.mittlerbros. com/mittler-bros/bead-rollers/power-drive.html). While some machines require a person to manually push and direct the metal sheet through the dies, other machines use motors to turn the dies that drive the sheet through, but still require a laborer to direct the orientation of the sheet. A critical difference between the existing manual processes and the one explored in the research is that in existing processes the machine is stationary and a sheet is moved through, whereas in our process the sheet is held in place and the tool is moved along the sheet. In manual processes the worker moving the sheet is constantly providing variable amounts of pressure at different points in the process to counter deformations. By clamping the sheet in place, deformations in the sheet are limited thereby allowing the robot to access the plane of the sheet as a constant reference.

3 Advances in Sheet Metal Fabrication Processes

The adaption of advanced CAD/CAM technologies by leading metal fabricators has opened up a new realm of potentials for sheet metal, which is being realized across multiple industries. Most applications of thin sheet surfaces are non-load-bearing, and tend to lend themselves towards compound curvatures. However, because of the molds and profiles that need to be created in cold-forming techniques, these processes are frequently quite cost prohibitive at a building scale (Schodek et al. 2005). If one were to remove the necessity for molds and secondary support systems, it may be possible to strengthen the economic argument for customized sheet metal components in building applications.

A number of precedents exist for the use of 6-axis industrial robots for various metal forming processes. One of the processes that has attracted much research is incremental sheet forming (ISF), a CNC die-less metal forming process that shapes metal by pressing a rotating spherical tool along pre-defined toolpaths.

The material may either be moved over a stationary forming tool or the tool may be moved while the material is fixed in place (Figs. 1, 2). Research out of TU Dortmund titled 'Robot Assisted Asymmetric Incremental Sheet Forming: Surface Quality and Path Planning' presented at Rob|Arch 2012 looked at Asymmetric Incremental Sheet Forming (AISF) as an architectural fabrication technique (Brüninghaus et al. 2012). The process outlined in this research shares similarities with robotic bead rolling in that it is a highly flexible process, and is best suited for small batches of customized parts. Another research direction with AISF is exploring the application of formed sandwich panels (Jackson et al. 2008). This application is relevant as it addresses certain geometric possibilities and potentials implied by the robotic bead rolling process. One of these geometric possibilities is the formation of patterns similar to that seen in an isogrid panel—a homogenous structure in which load-bearing ribs and the skin of the panel are milled from a single plate of aluminum alloy (Fig. 3). These types of panels have been developed by the aerospace industry for their strength and light weight, which are features that can be deployed for a variety of applications. While these panels deal with solid section corrugations involving a subtractive manufacturing process, shell corrugations—involving material expansion locally to increase overall depth— raise similar functional characteristics.

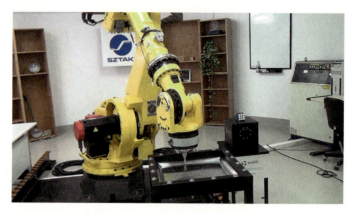

Fig. 1 Robotically   controlled   ISF   processes   (http://i1.ytimg.com/vi/xcg08U_Y4PI/maxresdefault.jpg)

Fig. 2 Robotically controlled ISF processes (Taleb Araghi et al. 2011)

Fig. 3 Image of lightweight isogrid panel (http://www.amrofab.com/)

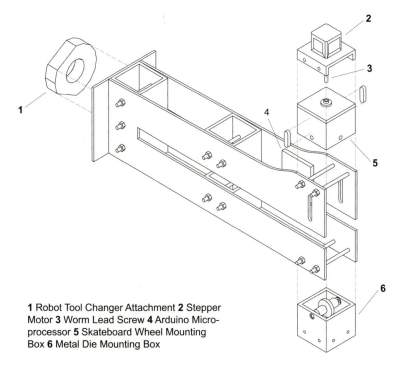

1 Robot Tool Changer Attachment 2 Stepper
Motor 3 Worm Lead Screw 4 Arduino Micro-
processor 5 Skateboard Wheel Mounting
Box 6 Metal Die Mounting Box

Fig. 4 Diagram of tool affixed to robot

4 Tooling and Process

4.1 Tool Development

With limited precedents surrounding variations of the traditional bead rolling machine, a trial and error approach was applied during the design of the tool (Fig. 4). Early prototypes in plywood reveal that there was a heavy amount of torque on the tool itself when a sheet is pushed through. Additionally these tests proved that the process requires both wheels—above and below the sheet—to spin in order to avoid scratching the surface of the sheet metal. To add more flexibility to the process a rubber skateboard wheel was tested in order to allow for variations in the bead depths. An added benefit of this process is that it allows for a bead being rolled into a sheet to intersect with existing beads without destroying the profile. This ability opens up opportunities for improvements in the structural performance of the sheet, as well as expands the lexicon of pattern possibilities. In later iterations of the tool, more attention was focused on how to provide varying depth to the beads, and exploring the possibilities of having rotating wheels. A lead screw mounted on top of the casing for the top wheel pushes down onto the casing, which drives the rubber wheel into the metal die. This process is now automated so

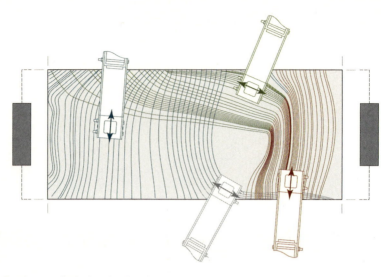

Fig. 5 Diagram displaying the directionality of the lines that are achieved with a 90° rotation of the wheel casings

that a Bluetooth device is connected to an Arduino that remotely controls the stepper motor that drives the lead screw into the wheel casing.

Various tests have been conducted to determine the benefits of allowing the wheels to freely rotate in order to allow the wheels to self-adjust to the movement of the robot in any direction. The mechanics of this process have proven to be somewhat unpredictable, and have caused errors due to the amount of force required to push down on the top wheel. As an alternative, the current tool is designed to allow the wheel to be mounted either parallel or perpendicular to the face of the tool, but without free rotation (Fig. 5). This flexibility in the tool was able to be sacrificed because of the flexibility granted by the movements of the 6-axis industrial robot. The latest iteration of the bead rolling tool provides a larger slot depth, and contains the remotely controlled motor that allows for exiting the material within the sheet.

In addition to the bead rolling tool developed for mounting to the robot. A jig has been constructed that holds the sheet in place during the robotic beading process (Fig. 6). By holding the sheet vertically as opposed to horizontally, the sheet stays flat. This is critical so that the tool can predict the location of the edge of the sheet.

Fig. 6 Jig ensuring sheet material is in place during the robotic beading

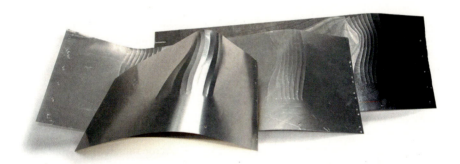

Fig. 7 Beading effect on sheets of different gauges

4.2 Material Testing

The majority of the material testing up to this point in the research has consisted of various gauges of aluminum sheets ranging from aluminum flashing to 20 gauge sheets, with 20 gauge being the most common due to its wide availability. This material is relatively cheap and malleable when compared to other stock metal sheets. 24 gauge mild steel has also been successfully tested during our prototypes. By altering the depth of the sheet through the process, further rigidity and strength are provided without the need of altering the chemical properties of the metals. While our prototypes have primarily used thinner and softer metals, at an industrial scale, thicker and stronger sheets would certainly be plausible for this process. Figure 7 illustrates the resulting sheet deformations from beads rolled in aluminum flashing, 20 gauge aluminum, and 24 gauge mild steel.

5 Robotic Workflow

5.1 Overview

Finite Element Analysis is performed on sheet metal samples given a set of user-defined load parameters and boundary conditions. The outcome is a series of stress vectors acting on the sample. These are hence identified as locations of weakness and where bead corrugations are introduced as means of enhancing the sample's structural performance towards a specific loading condition. Vector direction is translated into a toolpath and fed into the robotic controller while vector magnitude drives the corrugation depth through the tool-mounted microprocessor.

5.2 FE Analysis Results as Input

To demonstrate this workflow, Millipede, a Grasshopper® plugin developed by Sawako Kaijima and Panagiotis Michalatos, was explored as an analysis tool. Using a library of structural algorithms, Millipede performs Finite Element Analysis on linear elastic systems. It generates a set of statistical data simulating material behavior under different loading and boundary conditions. Basic assumptions are fed into the plugin including, geometry: planar material of constant thickness and density with dimensions that fit within the maximum workable domain of the operating tool; support regions: designating an area of structural support; and load regions: designating an area and vector value of load application. These parameters are ideally an output of a parametrically generated model allowing for quick prototyping of a wide range of conditions. An analysis model is thus generated in the form of a mesh which is then passed on to the Finite Element solver. While Millipede outputs several analysis types, the authors' interest lies first in principal stress vectors as curves travelling through the material, with either positive values for tension and negative for compression; and second—the Von Mises stress at points along these curves as a measure of combining both principal stress values (Figs. 8, 9). These vectors (curves) and their corresponding values simulate force lines used in visualization of internal forces in Solid Mechanics.

As the model is discretized into a smaller finite number of elements through meshing, internal surfaces to which the principal stress direction is perpendicular are identified. Numerical Integration of stress on each of these surfaces then allows for the visualization of overall distribution of stress patterns across the entire geometry. As the research has focused testing on planar sheets of material, the decision has been made to accept the marginal differences between stress and bending moment vectors for planar surfaces. If the analysis was to move into the curved surface realm, bending moment vector values would be of prime importance.

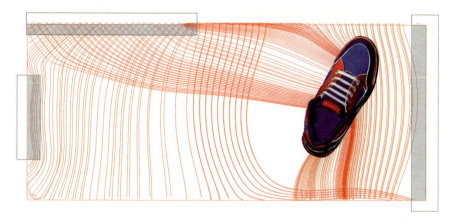

Fig. 8 Stress lines generated in millipede based on specific boundary conditions

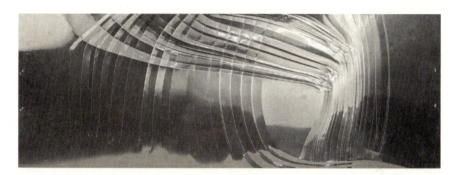

Fig. 9 Sample panel produced from millipede output curves

5.3 Digital Robotic Interface

With this data in hand, an interface processes the information into recognizable formats by other components down the workflow (Fig. 10). Principal stress curves are converted into toolpaths and used to feed HAL Robot Programming and Control—a plugin for Grasshopper® developed by Thibault Schwartz—eventually outputting the required RAPID code. Positive and negative values inform the corrugation directionality: outward corrugations resist tension while inward corrugations resist compression. As the majority of mechanical forces acting on a structural beam are concentrated at its upper and lower surfaces, the direction to which material is redistributed or corrugated becomes essential in resisting these forces. This is deemed particularly evident when testing two sandwiched planar metal sheets with tension on the upper sheet (outward corrugations) and compression on the lower (inward corrugation) (Fig. 11). The RAPID code is thus programmed in such a manner so as to allow tool disengagement from the

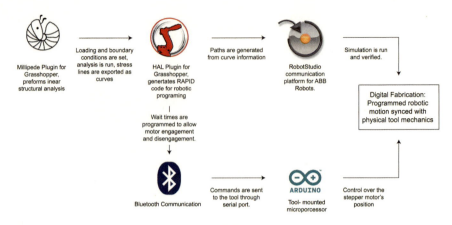

Fig. 10 Workflow for toolpath generation and motor control

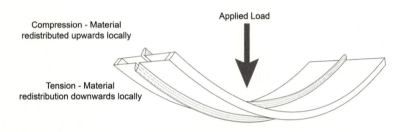

Fig. 11 Sandwiched planar metal sheets with bead corrugations

material, full 180° rotation, and re-engagement in order to flip corrugation directions.

On the other hand, Von Mises stress scalar values are remapped along a new distance domain in millimeters. These are supplied into Firefly, another Arduino control plugin for Grasshopper® developed by Andy Payne, allowing the tool-mounted stepper motor to travel accordingly. A higher stress value is translated into a larger motor pin displacement pushing the rubber wheel and metal die closer to one another and thus creating a deeper corrugation. Syncing the RAPID program with the Arduino component has proven to be challenging since the robot's travel time involves acceleration, deceleration and constant speed making it difficult to quantify. In current testing, the stepper motor is manually controlled via Bluetooth from a serial port while RAPID is being executed. The user is then able to control whether the two wheels are engaged (corrugations executed) or released (corrugations not executed), in addition to the corrugation depth. Future improvements include utilizing the ABB Robot Reference Interface as it allows for the periodic exchange of actual robot position data and other RAPID multitasking options through TCP. Such data is then received by the Arduino microprocessor prompting the execution of code and thus precise control. Within this closed loop system, robot movements would, in turn, be triggered upon feedback from the tool. This allows a fully automated, pre-programmed fabrication process.

5.4 Fabrication Environment

The jig to hold the metal sheets is designed in such manner so as to clamp the sheets along three sides with an exposed fourth side for tool access. Breaks are introduced in the RAPID to manually rotate the sheets. Careful consideration went into allowing safe distances between the clamps and the tool and avoiding collisions. A future iteration of the table could potentially involve a rotating base that allows automatic rotations of sides staying within the robot's maximum reach. Other iterations could support a number of sheets and allowing a larger set of prototypes to be produced within one robotic session. Over the course of prototyping, human intervention during the fabrication process has decreased with potential in reaching full automation as the research concludes.

6 Industrial Applications

6.1 FE Analysis Applications

Conventional use of FE structural analysis involves (a) The development of an initial model based on a structural engineer's experience or a designer's concept, and (b) Analyzing the model as means of providing a feedback loop informing what design decisions are to be made. The initial model is thus reiterated and continues to be analyzed until it has been optimized satisfying all design parameters. This iterative process includes both addition of structure to areas of high stress concentration and elimination at areas that lack stress forces. The input workflow outlined here aims at flattening this two-step process with its associated feedback loop into a single step procedure by using the FE analysis results as initial design criteria. By approaching material stock as an empty canvas and only treating areas that require additional strengthening, a smoother and simpler workflow is achieved. Additionally, in contrast to topology optimization that deals with altering material distribution, simplifying the core problem into 2D flat monomaterial studies allowed the authors to move away from the additive/subtractive nature of the existing workflow to a more transformative one. Parameters beyond structural performance including manufacturing constraints and heat flux amongst others could then all be added to early FE analysis. While these constraints are quantitative, varying their degrees of compliance could then allow for a set of design iterations—all within the acceptable quantitative threshold—on which

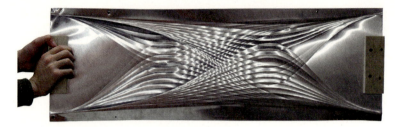

Fig. 12 20 gauge aluminum sample produced by robotic bead rolling process

qualitative aesthetic judgment can be made. This instigates new formal languages that arise out of pure functional and numerical investigations.

6.2 Robotic Bead Rolling Applications

Stressed skin surfaces, as stand-alone elements in monocoque arrangements or in tandem with structural members in composite systems, demonstrate a high performance capacity through their use in ship building and aerospace industries (Bechthold 2008). From building cladding panels to interior book shelves, stressed skin fabrication via robotic beading provokes a wide range of applications on various scales. This is due to the fact that it provides structural integrity to sheet metal and increases its rigidity. The authors, however, have recognized that these structural improvements do not reach the extents of using high gauge metal sheet as primary support in an assembly. Therefore, an added-value that comes with the application of such a technique would be allowing for lighter substructures resisting less mechanical forces. In the case of building facades, laminating multiple flat sheets that have gone through this strengthening process suggests that the desired visual flatness can be maintained while a structural layer on the underside supplies the rigidity.

While the authors have chosen to focus on analytical methodologies, bead rolling as an aesthetic impression and its associated play of light across corrugations is also a valid potential application of this process (Fig. 12). By varying the input mode beyond the suggested FE results, any algorithmically generated pattern could be easily formalized in the form of a rapid visual prototype. Bead corrugations in such prototypes do contribute to structural rigidity alongside pure aesthetics. This is due to the increase in the overall active structural depth, regardless of specific bead locations. Further directions of research would enable the 3d digital modeling of beads into a given geometry before production is carried out. This allows visual pre-verification as means of simulating fabricated outcomes. Assuming the reliability of digital testing matches its physical counterpart, this added function would also eliminate the need to physically test samples as further analysis, if required, could then be run on digital models.

7 Conclusion

Ongoing research has focused on digitizing certain elements of the traditional bead rolling process to allow for improved control and precision, while enabling novel formal possibilities that would have proved either difficult or impossible through manual processes. An automated process from digital form-creation to physical robotic fabrication has been developed entirely within the Rhinoceros environment. Toolpath input ranged from stress vectors produced through FE analysis to parametrically generated curve patterns. While both of these approaches added structural rigidity to the samples, further mechanical testing of samples is necessary to verify if there are significant structural differences between approaches. This type of analysis would allow the authors to then modify the manufacturing variables accordingly. Results from the panels produced throughout the research demonstrate that some appropriate applications include assemblies requiring load transfer through skins or lightweight sandwich panels. Additionally, the decorative possibilities enabled by the process have proven very provocative.

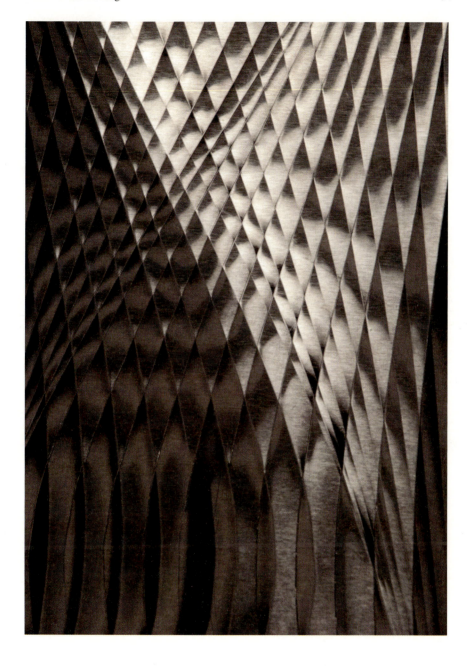

Acknowledgments This research was conducted under the guidance of instructor Andrew Witt and fabrication specialist Burton LeGeyt during the course: Expanded Mechanism/Empirical Materialisms at the Harvard University Graduate School of Design; Cambridge MA; Fall 2013.

References

Bechthold M (2008) Innovative surface structures: technologies and application. Taylor & Francis, New York, pp 189–191

Brüninghaus J, Krewet C, Kuhlenkötter B (2012) Robot assisted asymmetric incremental sheet forming. In: Brell-Cokcan S, Braumann J (eds) Rob|Arch Robotic Fabrication in Architecture, Art, and Design. Springer, Vienna

Jackson KP, Allwood JM, Landert M (2008) Incremental forming of sandwich panels. J Mater Process Technol 204(1–3):290–303

Schodek D, Bechthold M, Griggs K, Kao KM, Steingberg M (2005) Digital design and manufacturing: CAD/CAM applications in architecture and design. Wiley, Hoboken, pp 59–61

Taleb Araghi B, Göttmann A, Bambach M, Hirt G, Bergweiler G, Diettrich J, Steiners M, Saeed-Akbari A (2011) Review on the development of a hybrid incremental sheet forming system for small batch sizes and individualized production. Prod Eng Res Dev 5(4):393–404

A Compound Arm Approach
to Digital Construction

A Mobile Large-Scale Platform for On-site
Sensing, Design, and Digital Fabrication

Steven Keating, Nathan A. Spielberg, John Klein and Neri Oxman

Abstract We introduce a novel large-scale Digital Construction Platform (DCP) for on-site sensing, analysis, and fabrication. The DCP is an in-progress research project consisting of a compound robotic arm system comprised of a 5-axis Altec hydraulic mobile boom arm attached to a 6-axis KUKA robotic arm. Akin to the biological model of human shoulder and hand this compound system utilizes the large boom arm for gross positioning and the small robotic arm for fine positioning and oscillation correction respectively. The platform is based on a fully mobile truck vehicle with a working reach diameter of over 80 feet. It can handle a 1,500 lb lift capacity and a 20 lb manipulation capacity. We report on the progress of the DCP and speculate on potential applications including fabrication of non-standard architectural forms, integration of real-time on-site sensing data, improvements in construction efficiency, enhanced resolution, lower error rates, and increased safety. We report on a case study for platform demonstration through large-scale 3D printing of insulative formwork for castable structures. We discuss benefits and potential future applications.

Keywords Digital fabrication · Robotics · Large-scale fabrication · 3D printing · Insulated formwork · Digital construction platform (DCP)

S. Keating (✉) · N. A. Spielberg · J. Klein · N. Oxman
Massachusetts Institute of Technology, Cambridge, USA
e-mail: stevenk@mit.edu

N. A. Spielberg
e-mail: nspielbe@mit.edu

J. Klein
e-mail: j_klein@mit.edu

N. Oxman
e-mail: neri@mit.edu

W. McGee and M. Ponce de Leon (eds.), *Robotic Fabrication in Architecture, Art and Design 2014*, DOI: 10.1007/978-3-319-04663-1_7,
© Springer International Publishing Switzerland 2014

1 Introduction

Robotic systems and digital fabrication technologies are increasingly implemented for architectural design and construction through a variety of material processing techniques and delivery systems (Bonwetsch et al. 2006; Gramazio and Kohler 2008; Lim et al. 2012). From brick-laying robotic arms (Bonwetsch et al. 2007) to filament winding of structural composite parts for architectural construction (Castro et al. 1993; Shirinzadeh et al. 2004), robotic construction for large scale systems is enabling architects and engineers alike to further push the envelope of digital design generation and automation. Furthermore, such construction technologies can serve to fundamentally challenge and transform well practiced design traditions.

Despite this promise, on-site design, sensing, and fabrication for large structures are still lacking. In their absence, architects are forced to segregate between processes of design generation and processes of design construction. Furthermore, the deployment of robotic fabrication tools and technologies is typically limited to the production of prototypes or off-site small-scale structures.

In this chapter we present research-in-progress for a novel digital platform capable of on-site design, sensing, and fabrication of large-scale structures. The system combines a large hydraulic boom arm and a smaller electric robotic arm. Through the control of both arms, the system enables digital fabrication processes at architectural scales. As a result, the Digital Construction Platform (DCP) (Fig. 1) opens up new opportunities for on-site sensing, design, and fabrication capabilities.

By attaching a sensor as an end effector to the system, it is able to generate volumetric scanning of physical geometry and environmental conditions, such as radiation, soil stability, temperature, and chemical mapping (Keating and Oxman 2012). Such scanning technologies have been previously explored as stand-alone applications (Callieri et al. 2004). However, by combining sensor data with material deposition logic the designer is able to respond to site-specific environmental data and terrain mapping in real time. In addition, various fabrication methods using a range of end effectors such as a mill, an assembly gripper, or a welding tool, can be implemented as part of the fabrication process. Through these capabilities the DCP system is designed to complete the large-scale tool chain, providing real-time digital sensing, on-the-fly performance-based design, and on-site construction. Potential benefits include the construction of complex architectural forms, improvements in efficiency, enhanced resolution, lower error rates, recorded measurement data, and significant increases in safety.

We demonstrate the use of large-scale 3D printing of formwork utilizing the Digital Construction Platform as a first case study implementation. Using quick cure polyurethane foam, insulative formwork can be rapidly 3D printed in doubly curved geometry without the need for support material. Early analyses of foam strength, print speeds, resolution, and preliminary economic figures confirm the feasibility of the project (Keating 2012).

Fig. 1 The digital construction platform (DCP) is comprised of a 6-axis KUKA robotic arm mounted to a 5-axis Altec hydraulic boom arm

2 Platform Design

The Digital Construction Platform utilizes a mobile system capable of an extended physical reach, complex functional movement and high load capacity thereby enabling new modes of in situ construction. The design of the platform was motivated by the need for a flexible system capable of implementing various kinds of large-scale digital fabrication approaches including additive, subtractive, and assembly techniques. Utilizing an extended stationary reach, the DCP system can handle large load capacities as well as high degrees of access and accuracy (Fig. 2).

Furthermore, the DCP functions as a mobile system allowing for fast setup times and ease of repositioning. Compared with existing construction platforms, hydraulic boom arms prevail for flexible manipulation from a stationary position (Saulters and Scarr 1985). However, these systems typically lack the precision or automation required for digital fabrication techniques. The DCP is designed around a hydraulic boom arm with an added robotic arm effector for compensation of oscillations, increased precision, and ease of access.

The current platform utilizes a GMC truck, an Altec L42M boom arm, and a KUKA robotic arm, providing a lift capacity of 1,500 lb (boom arm mount) with a manipulation capacity of 20 lb (KUKA arm). The 6-axis KUKA robotic arm is mounted on the end of a two-axis hydraulic jib on the 3-axis boom arm. The system uses a KUKA KR5 sixx R850 arm and is currently being upgraded to a KUKA KR10 R1100 arm for improved mechanical specifications. The robotic arm is controlled via a custom python script package enabling real-time control via the

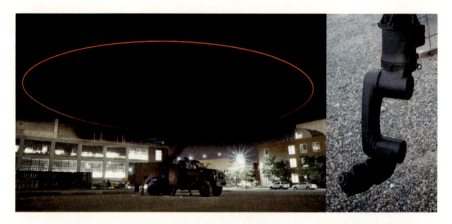

Fig. 2 The range of motion for the DCP is shown through long-exposure photography (*left*). The DCP uses a combination system of a 5-axis Altec hydraulic boom arm and a small 6-axis KUKA robotic arm (*right*)

KUKA Robot Sensor Interface (RSI). The controls system is designed as feedback loop based on real-time data from magnetostrictive sensors, rotary encoders, and inertial measurement units.

3 Control System

The control system is designed to accomplish three goals: accurate positioning, high speed, and minimization of oscillations at the end effector. Due to its can-tilevering structure, slow persisting oscillations in the boom arm are encountered at the end during open loop control as seen in Fig. 3 and are common in con-struction equipment (Yong et al. 2005). The oscillation amplitude and duration can be reduced through a controls scheme. However, we must still compensate for the dynamic response to gain the desired positional accuracy and speed for digital fabrication applications. We aim to operate at end effector speeds of up to 0.2 m/s and a targeted accuracy of ± 0.5 cm. In order to achieve this desired speed and accuracy we plan to incorporate a linear quadratic controller designed to minimize the magnitude of oscillations at the end effector of the large boom arm. Besides this controller, we plan on also implementing a position control scheme to control the localization of the end effector relative to the boom arm.

In order to sense such oscillations, we use the ADXL345 accelerometer and the ITG-3200 MEMS gyroscope on the end of each cantilevered linkage. The velocity of each hydraulic piston on the boom is controlled by mechanically actuating the hydraulic valves with servos. Each hydraulic piston is treated as an ideal velocity source in the control system model. Joint angles are sensed with Balluff BTL6 Micropulse transducer sensors for measuring hydraulic extension. At the base of the arm, the rotational angle is sensed by using a YUMO E6B2 encoder mounted

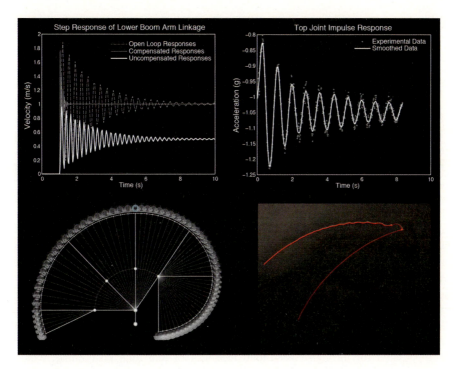

Fig. 3 The control model for the DCP allows a range of motion (*bottom left*) and compensates for robotic arm oscillations (image *bottom right*, acceleration data *top right*). The figure demonstrates a simulation of open loop, compensated and uncompensated response from the controls model (*top left*)

on the rotational hydraulic motor. Each of the sensors is designed to provide feedback for the end effector position through the kinematics. Furthermore, global positioning feedback will be established through the use of a ground reference sensor. The use of LIDAR and other environmental sensors such as magnetometers and ground penetrating RADAR could also provide three-dimensional environmental data for use in optimal construction of structures.

In the current controls model, by linearizing the boom linkage joints about a small angle and deriving a system model, we demonstrate that a PID controller is able to reduce the time and amplitude of open loop system oscillations (Nise 2011). The system is linearized about a nominal optimal operating position. In small deviations from this nominal position the linearized model is valid within acceptable error. We are currently investigating a nonlinear system model for improved response.

By using a STMicroelectronics LIS331DLH MEMS 3-axis digital accelerometer with a sensitivity of 1 mg/digit, we have characterized open loop system properties. This was achieved by performing an impulse response test in each joint of the boom arm (Fig. 3). From this impulse test we derive the mass, spring, and damping system characteristics by quantifying the period and decay of the measured acceleration output. These mass, spring, and damping characteristics dictate the open loop system dynamics and are used in simulation to construct an estimated system model. The simulation results conducted in MATLAB Simulink for the boom arm are given in Fig. 3. In order to compensate for these minimized oscillations, the KUKA robotic arm operates via a custom python script using the RSI package for real-time control. The python script enables the operation of the KUKA controller via packet communication from a main computer that integrates the programmed tool path, sensor data, and a graphic user interface. The KUKA motions are dictated through feedback from an accelerometer at the base of its arm that enables compensatory real time trajectories. The resulting compensation action is made possible by the fast response time of the small robot arm in comparison to the frequency of the boom oscillations. The home location of the robot arm end effector is designed to be located in the center of its spherical working area. This home position will allow for maximum oscillation amplitude adjustment. The two systems work in tandem through a unified controls scheme, with the addition of feedback from each of the joints.

We intend to incorporate additional sensors to provide a ground reference point for redundancy. In addition, this secondary ground reference sensing will reduce hysteresis and small error accumulation in the position measurement of the end effector. Other errors such as the change in external working temperature will be compensated for through closed-loop feedback of real time positioning, which increases the robustness of the controls system. Interference can be minimalized by the use of a real time kinematic GPS sensing system for ground referencing. The combination of this ground reference sensor along with the other accelerometers, and position sensors on each joint will close multiple loops allowing for accurate end effector positioning.

4 Material Case Study

Robotic fabrication systems provide increased efficiencies of existing processes, enable transfer of unsafe human techniques to machines, support novel manufacturing techniques, and—most importantly—open new design possibilities for a more integrated architecture. The DCP is not intended to replace people; rather, it is designed to augment capacities in which robotics excel, namely precise, repetitive, and dangerous tasks. In addition, the DCP is designed to allow for novel on-site integration of sensor data, design, and fabrication techniques. Instead of a static design, the DCP supports data capturing and integration of site data such as topography, material analysis, and environmental conditions. This concept of real-time sensing and integration into fabrication processes offers improvements in form, reliability, and customization of design form to fit functional and site-specific environmental requirements. Under this framework we introduce a novel additive fabrication technique enabled by robotics as a case study implementation for our DCP system.

Since the early days of assembly lines, multiple proposals for automated housing construction have been attempted. Thomas Edison's single-pour concrete housing, patented in 1917 (Edison 1917), was one of the first to explore this concept. Edison's dream of a re-useable mold for castable concrete houses, complete with fine details, resulted in a well-documented failure due to mold issues and the inability to customize individual designs. In modern day practice, housing construction techniques are primarily based on manually assembled structures while automation holds significant potential to improve safety, form, and efficiency. The current field of architectural-scale additive manufacturing has explored direct extrusion of cementitious materials or the use of binder material, as seen in Contour Crafting or D-Shape (Khoshnevis 2004; Khoshnevis et al. 2006). However, limitations relating to materials, complex forms, and on-site scalability still remain (Lim et al. 2012).

To enable the variation of material properties with any castable material while providing enhanced speed, a new technique based on formwork was created. Similar to insulated concrete forms, leave-in-place insulating formwork can be 3D printed for castable structures. The process, termed Print-in-Place construction, is designed for on-site fabrication of formwork for castable structures, such as concrete exterior walls and civil infrastructure (Keating and Oxman 2013). Print-in-Place Construction provides a method by which to overcome the complexities associated with direct concrete extrusion. In addition, mold printing allows for a stronger product as the material is instantly cast instead of it being constructed through successively layering (Keating and Oxman 2013).

The polyurethane spray foam utilized in the prototype system (Dow Chemical FROTH- PAK foam) is a two-component chemical foam with a cure time of 30 s. It is an expanding foam that is strong, lightweight, and designed for a high insulative value. Due to the rapid cure time, a large structure can be printed very quickly. For example, the curved twelve foot long wall structure demonstrated in Fig. 4 was printed in less than 5 min using the KUKA robotic arm to control the nozzle position. Time estimates for small exterior house structures are under a day (Keating 2012).

In addition to additive printing, the Print-in-Place technology utilizes subtractive techniques to improve surface finish and reduce manufacturing time. The formwork is printed in thick layers of around two inches enabling fast build times followed by a surface mill to achieve a higher resolution. The resulting resolution from a cast structure inside a printed and milled mold is shown in Fig. 5.

Furthermore, subtractive processes combined with embedding objects (such as rebar or tie structures) in the printing process enable achieving complex details such as windows, wiring areas, and embedded sensor integration.

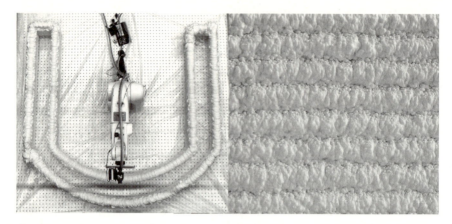

Fig. 4 Additive fabrication tests using polyurethane spray foam with a KUKA 6-axis arm (*left*) produced test insulative formwork samples with consistent and tunable layer heights (*right*)

Fig. 5 Combining additive and subtractive processes in a compound end effector (*right*) facilitates fast build times and high resolutions, as seen in the cast structure produced from a printed and milled mold

Compared with traditional construction practices, the benefits of the designed system are substantial. Whereas the former method requires human labor and large construction machines, the latter allows for buildings to be printed with cost-effective mobile printing units. From a material standpoint, the DCP system wastes few resources and uses only the amount of bulk material that is required for construction. This reduces the price of construction down to the bare minimum based on the price of the bulk material components (insulating foam and castable materials for structure). From a speed standpoint, reducing the construction site time and integrating processes improves efficiencies. Finally, custom formal manipulations are easily achieved, as the geometry is not constrained by rectilinear paths.

Overall, printing formwork is an effective method that offers benefits over the difficulties involved in printing concrete directly or using a powder/binder process. By printing dual-purpose foam that acts as a mold for the concrete as well as insulation for the building, Print-in-Place Construction is significantly more versatile and can incorporate different materials or variations of concrete. The process can also be rapidly integrated into current building strategies and regulations as the Print-in-Place Construction method aligns directly with traditional insulated concrete form (ICF) technology. Once the mold is printed, conventional methods and regulations that apply to ICF construction are applicable to the Print-in-Place process (Keating and Oxman 2013).

5 Conclusions

The Digital Construction Platform (DCP) is designed as an enabling design and construction platform to sense, design, and construct highly integrated architectural constructs. The increased capability to automate synchronized fabrication sequences creates the opportunity to design an interwoven set of relationships between structural, architectural, and environmental systems—in turn enabling true building integration. This is achieved through the DCP's versatility and capacity to create feedback loops between real-time site-specific sensing and fabrication processes. Compound arm techniques on a mobile platform bridge the digital and physical domains by enabling the transition of digital fabrication into large-scale on-site building construction. Our investigations into additively fabricated insulative formwork highlights the type of novel possibilities enabled by the DCP platform. Future work entails fitting out the mechanical and sensing systems, completing material testing and investigating multi-platform collaboration with swarm construction techniques. Finally, we aim to design and construct a full-scale architectural pavilion using the DCP system in the near future.

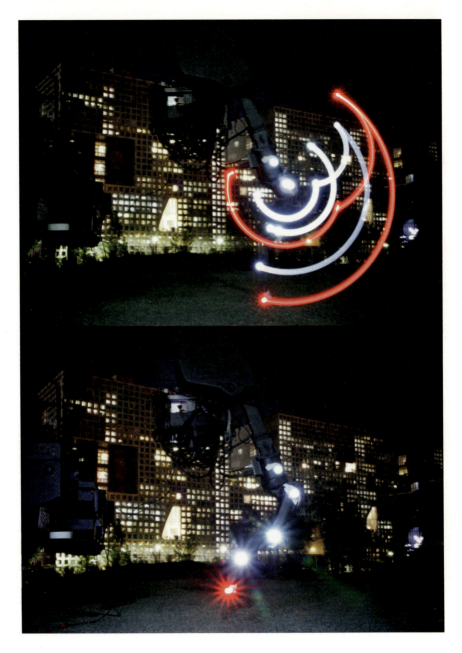

Acknowledgments We would like to acknowledge Altec Industries for their support of this project and technical assistance. In addition, this research was supported in part by an NSF EAGER grant award #1152550 "Bio-Beams: Functionally Graded Rapid Design and Fabrication". We would like to thank the Mediated Matter group and the MIT Media Lab for their kind help and support. Specifically, we would like to acknowledge Ali AlShehab, Benjamin Jennet, Dr. David Wallace, Dr. Franz Hover, Hannah Barrett, Sergio Falcon, Will Bosworth and Xiaoyue Xie for their assistance with this project.

References

Bonwetsch T, Gramazio F et al (2007) Digitally fabricating non-standardised brick walls. In: Proceedings of ManuBuild, 1st international conference, Rotterdam

Bonwetsch T, Kobel D et al (2006) The informed wall: applying additive digital fabrication techniques on architecture. In: Proceedings of ACADIA, Louisville, Kentucky

Callieri M, Fasano A et al (2004) RoboScan: an automatic system for accurate and unattended 3D scanning. In: IEEE 2nd international symposium on 3D data processing, visualization and transmission

Castro E, Seereeram S et al (1993) A real-time computer controller for a Robotic Filament Winding system. J Intell Rob Syst 7(1):73–93

Edison T (1917) Apparatus for the production of concrete structures. Patent No. 1326854

Gramazio F, Kohler M (2008) Digital materiality in architecture. Lars Muller Publishers, Baden

Keating S (2012) Renaissance robotics: novel applications of multipurpose robotic arms spanning design fabrication, utility, and art. M.Sc. thesis, Mechanical Engineering, Massachusetts Institute of Technology

Keating S, Oxman N (2012) Immaterial robotic fabrication. In: Proceedings of RobArch: robotic fabrication in architecture, art and design

Keating S, Oxman N (2013) Compound fabrication: a multi-functional robotic platform for digital design and fabrication. Robot Comput Integr Manuf 29(6):439–448

Khoshnevis B (2004) Automated construction by contour crafting—related robotics and information technologies. Autom Constr 13(1):5–19

Khoshnevis B, Hwang D et al (2006) Mega-scale fabrication by contour crafting. Int J Ind Syst Eng 1(3):301–320

Lim S et al (2012) Development in construction-scale additive manufacturing processes. Autom Constr 21(1):262–268

Nise N (2011) Control systems engineering. Wiley, Hoboken

Saulters JD, Scarr RH (1985) Method and apparatus for controlling the position of a hydraulic boom. Patent No. 4514796A

Shirinzadeh B, Alici G et al (2004) Fabrication process of open surfaces by robotic fibre placement. Robot Comput Integr Manuf 20(1):17–28

Yong Y, Rydberg K, An L (2005) An improved pi control for active damping of a hydraulic crane boom system. In: Proceedings from the international society for optical engineering

Design of Robotic Fabricated High Rises

Integrating Robotic Fabrication in a Design Studio

Michael Budig, Willi Viktor Lauer, Raffael Petrovic and Jason Lim

Abstract Despite a growing interest in robotic fabrication in academic research, its impact on the design of large-scale architectural typologies has not yet been explored. At the Future Cities Laboratory in Singapore robotic fabrication was integrated in a design research studio to produce 1:50 scale models of mixed-use high-rise typologies. Its methodology aimed for the reconsideration of the traditional architectural model by directly linking the digital design process with physical manufacturing processes and tools. As such, it established a strong correlation between computational tools, material systems and robotic fabrication strategies. Since high-rises are strongly rooted in the industrialization of building, with repetitive elements stacked along the vertical axis, they represent an interesting architectural typology to be challenged by a new fabrication paradigm (Gramazio and Kohler 2008: Digital materiality in architecture. Lars Müller Publishers, Baden). This chapter presents customized robotic fabrication processes and tools that were developed in the 2 year long design research studio.

Keywords Robotic fabrication · High-rise · Architectural model · Design studio · Computation · Singapore

M. Budig (✉) · W. V. Lauer · R. Petrovic · J. Lim
Module for Architecture and Digital Fabrication, Future Cities Laboratory, Singapore-ETH Centre for Global Environmental Sustainability (SEC), Singapore, Singapore
e-mail: budig@arch.ethz.ch

W. V. Lauer
e-mail: lauer@arch.ethz.ch

R. Petrovic
e-mail: petrovic@arch.ethz.ch

J. Lim
e-mail: lim@arch.ethz.ch

W. McGee and M. Ponce de Leon (eds.), *Robotic Fabrication in Architecture, Art and Design 2014*, DOI: 10.1007/978-3-319-04663-1_8,
© Springer International Publishing Switzerland 2014

High-rises dominate large parts of the urban landscapes in fast growing regions throughout Asia. In cities like Singapore a majority of the population lives in residential high-rises.[1] The construction of this typology is strongly rooted in an industrial paradigm (Cousineau and Miura 1998). It is mainly driven by efficiency and economic criteria, with repetitive elements being stacked along a vertical axis. The questions arise, how contemporary computer aided architectural design with the integration of robotic fabrication could contribute to a more differentiated articulation and leverage more variety in the formal expression and functional capabilities of this widespread typology (Willmann et al. 2014). The design methodology itself comes into the focal point of investigations, which is tested in the context of a design studio.

The design research studio for master students is set up at the Future Cities Laboratory (FCL) in Singapore[2] in order to put significant research into this question. Through the robotic fabrication of 1:50 scale models of mixed-use residential high rises the experimental research studio investigates potential impacts of these technologies on the design and construction of novel high-rise typologies (Fig. 1). The studio is in constant exchange with PhD researchers in the robotic laboratory, which thus serves as an experimental test bed for both digital design and fabrication research (Hack et al. 2013). PhD research on computational design processes and the development of software environments to control robots plays a crucial role in the studio, and it offers in return test cases (Lim et al. 2013).

Within the design research studio teams of two to four students develop their architectural concepts based on the integration of computational design strategies and customized robotic fabrication processes (Fig. 2). The physical and the digital models are in constant negotiation with one another (Sheil 2005). Therefore constraints of the actually built model, e.g. in terms of material properties or manageable element dimensions, directly influence the computational design setup in a continuous feedback loop. The model scale of 1:50 requires a careful selection and abstraction of investigated aspects, but also demands a rigorous consideration of its tectonic logic (Budig et al. 2014). Up to four meter high models create their own construction reality. They oblige students to tackle problems of structural stability and the construction process from the very beginning on. This chapter discusses (a) the customized robotic system, (b) the embedded mechanical tools, like the development of end-effectors, and (c) it explains the research through case studies by illustrating how they conceived physical processes for the construction of the 1:50 models.

---

[1] In Singapore more than 80 % of the population lives in high-rise and high-density flats built by the Housing Development Board (HDB).

[2] The Future Cities Laboratory is run at the Singapore-ETH Centre for Global Environmental Sustainability (SEC), co-funded by the Singapore National Research Foundation (NRF) and the ETH Zurich. The design studio was run over two consecutive years, with a one-year program in each in 2012 and 2013. Participating students were from the Swiss Federal Institute of Technology Zurich (ETH) and National University of Singapore (NUS).

Fig. 1 Studio filled with series of tower models that were built in several iterations (Gramazio and Kohler, FCL Singapore, 2012, image by Callaghan Walsh)

Fig. 2 Image of robotic facilities with the final generation of tower models (Gramazio and Kohler, FCL Singapore, 2013, image by Callaghan Walsh)

1 Robotic System

Students and researchers share three customized robotic units. Each one consists of a small Universal Robots UR5 robot arm with six degrees of freedom that is mounted to an automatically driven Guedel axis configuration (Fig. 3).[3] This robotic system enlarges the working space of the robot arm from a range of 85 cm to an envelope of 4 m height, 1.7 m diameter and 2.7 m depth; due to its small operating diameter the robot arm can still reach very intricate locations. The setup is planned for the digitally controlled assembly of complex physical models within this working envelope (Brayer 2013).

The robotic towers are modularized to enable reconfiguration within the research laboratory. A high degree of modularity accommodates quick modifications of the robotic system, e.g. their height and thus the operating space can be adjusted without the need of any special tools. Four adjustable base points allow the robotic tower to be leveled and transfer its 1.2 t to the floor. This configuration has two advantages: First, it causes little vibration, so no additional floor enforcement is needed. Second, it ensures an equal distribution of dynamic and static loads, which is crucial on floor constructions without specifications for machine installation. Hence the robots can be taken out of the typical industrial environment and implemented in a design studio environment (Fig. 4).

2 Physical Tools and End-Effectors

For the development of a robotic fabrication process in the design studio the robotic end-effectors become the most crucial physical components. Available mechanical grippers are mostly not flexible enough to grasp pieces of various sizes and geometries, and can hardly be adapted to different assembly concepts (Kripper et al. 1989). To overcome any limitation a modular gripper system was developed to enable multiple options of mechanical and vacuum suction gripping. Students can produce these grippers easily and develop their own configurations, in order to figure out the solutions for the model building process (Fig. 5). While the initial focus was put on the gripper geometries for the control of picking and placing behavior, special concepts were eventually designed with more functional integration, such as sensor equipment and high-resolution control valves for optimized vacuum suction grippers (Monkman et al. 2007).

Since the previously developed grippers with suction cups restricted the building components' geometries, a second generation of vacuum grippers emerged from the idea to design a surface with hundreds of small apertures. The gripper is built out of three layers. The first layer is the perforated surface, with the air-feeding layer below and the third layer covering the feeding cavity from the

---

[3] Universal Robots UR5 robot arms are integrated in Guedel 2-Axes Linear Modules Type ZP-3.

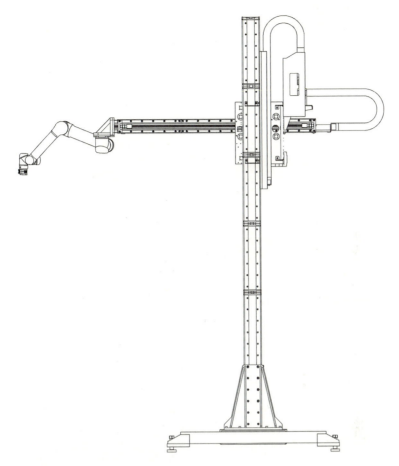

Fig. 3 Elevation of one of the robotic fabrication units; a Universal Robot UR5 robotic arm is mounted to a Guedel axis system in order to increase the building envelope (Gramazio and Kohler, FCL Singapore, 2012)

backside. With this configuration grippers could easily be produced with a laser cutter and optimized for the elements' geometries. Due to the thin buildup of the grippers of 3–5 mm, they were better suited for dense assemblies at 1:50 model scale (Figs. 6 and 7).

Unlike their industrial counterparts the lightweight Universal Robots do not need to be sheltered in a safety environment. The UR5 robot arm is equipped with built-in safety systems that allow its use without any further safety measurements. This opens the possibility for direct human–robot interaction, which became one of the primary advantages of the robotic facilities for a continuous refinement and adaption of robotic end-effectors. Only when the whole system, including the horizontal and vertical Guedel axes, are in full operation, a higher-level security

Fig. 4 A simple pulley is used for installation and height adjustments; the vertical axis is modularized into five segments (Gramazio and Kohler, FCL Singapore, 2012)

installation comes becomes a requirement. This is met with a laser scanner system, which registers any changes or approaches within a defined safety envelope. Users can still approach the facilities while the UR is in motion, but the Guedel axes will decrease their speed first and stop instantaneously on closer proximity. As such, the operating paradigm of the robotic towers is to combine the highest possible level of accessibility, human intervention and safety in the laboratory environment.

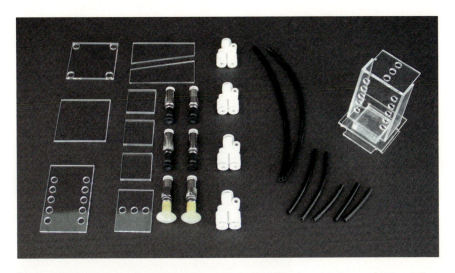

Fig. 5 Picture of a basic modular gripper setup that can be altered and amended by students (Gramazio and Kohler, FCL Singapore, 2012)

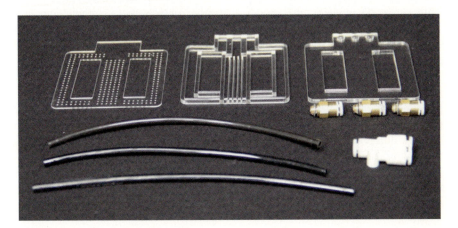

Fig. 6 A further griper development is constructed with three layers: perforated surface, and air-feeding layer below and the covering layer (Gramazio and Kohler, FCL Singapore, 2012)

3 Software Tools

In a similar vein to the customizable hardware components, a custom robot programming library called YOUR and a corresponding toolkit of Grasshopper components were developed that are open to end-user modification. These software tools aim to make robot control accessible to students without prior specific knowledge or programming skills. Students use the toolkit either from

Fig. 7 The thin buildup of
the gripper makes it suitable
for dense assembly
configurations (Gramazio and
Kohler, FCL Singapore,
2012)

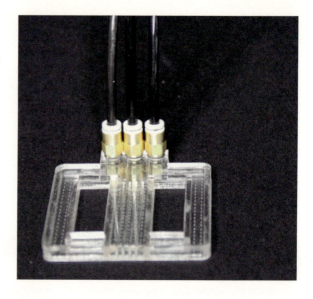

Grasshopper visual programming environment or from the script editor in the
Rhinoceros 3D modeler; the former is geared towards those without any programming experience while the latter suits experienced programmers. In either
case, students are able to control the robot directly from their computational design
environment.

By directly assembling components from the Grasshopper toolkit, students were
able to set up and control their robotic fabrication sequences. This visual programming approach facilitated students in quickly prototyping processes, as they
only needed to learn how a few essential components worked and could connect
them in different ways. Since the text-based code defining these components is
accessible, students become able to modify them once they acquire more experience in programming and knowledge in robotics. This allows them to introduce
more complex assembly logics and more intricate robot motion patterns for
material manipulation.[4]

4 Case Studies: Robotic Fabrication Processes

4.1 Picking and Placing

The first aim of the design research studio is to build models as high as possible to
gauge the limits of the robotic facilities. Initial towers are stacked configurations
that are fabricated in a simple pick and place fabrication process, for which the

---

[4] See Lim et al. (2013) for more details on the software environments.

students develop different vacuum gripper systems to glue and place cardboard elements. Hence in the beginning developments focus on the use of multi-functional grippers. They incorporate the gripping of pieces from different directions and have to consider material thicknesses, drying times, and height deviations caused by the applied layers of glue. The major challenge is to master the negotiation of a computer generated 3D model in a physical reality.

After this initial phase further strategies to utilize the robots' inherent manufacturing potential emerge and the concepts that become guiding principles for their tower designs. One concept in the studio investigates in an increased resolution of their towers by building with small components. This development culminates in the incorporation of a material feeder and a gluing device into a multi-functional end effector. Using spray glue, this system can hold several hundred pieces at a time and consequently speeds up the construction process by minimizing the distance the robotic arm had to travel for placing each piece. One of the resulting towers consists out of more than 15,000 pieces based on two different types. The focus shifts on structural systems that can cantilever outwards from the main vertical structural system (Fig. 8).

Another fabrication concept focuses on the smooth integration of the laser cutter, which allows students to produce elements with different sizes and geometries in an efficient workflow. Since the picking point varies for each piece, a corresponding feeder system and Grasshopper setup are developed. The cardboard containing the prefabricated elements gets constrained to fit into the robot's workspace. The data for the laser cutter is automatically generated in the digital model. The individual sheets are then placed directly on the feeder that contains a gluing station. Since the laser cut sheets are generated by code, the geometric data is used to coordinate the picking, gluing and placing movements of the robot. Although the process needs high precision, it enables fabrication processes with individual components and hence becomes a very generic and geometrically widely applicable process (Fig. 9).

4.2 Material Deformation Processes

Beyond picking and placing strategies the integration of a material deformation process explores the potential of the robot in its unique capacity to produce bespoke parts by just deforming identical elements. Taking the cardboard elements from previous studies a setup is developed that enables the bending of identical sheets at various angles. In a further iteration aluminum elements can be bent more precisely to defined angles. This leads to the production of large numbers of uniquely folded configurations out of the same generic element. By bending each piece in two opposite directions each wall stabilizes itself. Afterwards these walls are arranged in a way that they intersect with a wall on the story beneath to ensure continuous vertical load transfers.

Fig. 8 Customized gripper system that glues and places small building components in one step (Gramazio and Kohler, FCL Singapore, 2012)

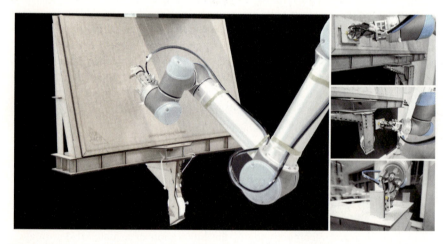

Fig. 9 Picture of a combined feeder and end-effector system; the feeder allows the students to directly place prepared laser cut sheets for the assembly process (Gramazio and Kohler, FCL Singapore, 2012, student project by Genhart P, Goldener P, Thonney F and Wullschleger T)

For the controlled bending process two pneumatically powered actuators are integrated in the robotic system. One mechanical gripper is holding the piece in place, while the other mechanical gripper is mounted on the robot arm. Through rotating this gripper around the stationary one, identical basic components are folded into unique parts (Fig. 10).

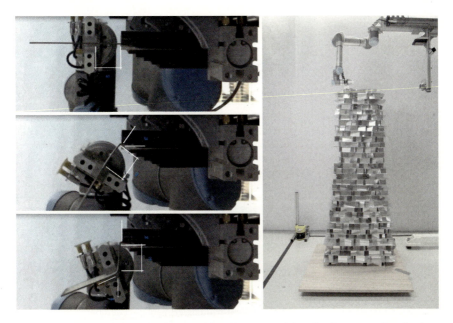

Fig. 10 Picture of cardboard folding process and tower model made of more precise aluminum sheets (Gramazio and Kohler, FCL Singapore, 2012, student project by Ernst S, Rickhoff S, Strohbach S and Tessarz M)

It proves to be a promising strategy to exploit the robot's unrestricted potential of spatial movements and by directly informing the material. Working with identical basic elements can eliminate logistic problems that are for example inherent in using the laser cutter in combination with the robot. In contrast to simple picking and placing, the robot plays an active role in the form-giving process. The deformation leads to the implementation of new sets of parameters related to material properties, for example a translation of a heating process of acrylic elements into the geometric data set of the modeling program (Grasshopper is primarily used for the robot control).

Two case studies integrate a heat gun in their robotic setups to deform plastic material, which allows students to locally heat material and then bend it by using the robotic arm. One concept builds up on the experiences with a high number of building elements (see project in Fig. 8) and in this further development involves bending acrylic stripes at multiple points to create a tower's primary structural system. After the thermal deformation, the pieces are cooled in order to avoid retraction and increase assembly speed (Fig. 11). A similar process is used in another case study for twisting acrylic sheets and producing a façade louver system. To expand the previously developed picking and placing process with the material deformation (compare with the process illustrated in Fig. 9), a combined mechanical and vacuum suction gripper was used (Fig. 12).

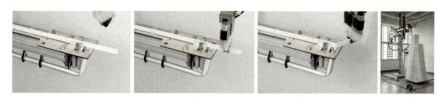

Fig. 11 The acrylic stripes are fixed in a linear rail and pulled forward to their designated bending position. The material gets heated up for 10 s and allows the robotic arm to bend the material to any angle between 0° and approximately 160°. After the deformation the material gets cooled down with air pressure (Gramazio and Kohler, FCL Singapore, 2012, student project by Kramer S and Stuenzi M)

Fig. 12 Combined gripper for picking identical building elements and deforming them with an integrated heat gun (Gramazio and Kohler, FCL Singapore, 2012, student project by Genhart P and Wullschleger T)

An entirely different approach to deform material is showcased by another case study with robotically stapled paper strips. This process requires the development of a gripper that can pinch two stripes of paper and then staple them together to fixate their positions. By sliding the two paper stripes with different intervals and stapling them again it is possible to produce geometries with undulating outlines (Fig. 13). The resulting geometries are used to produce layers of exterior enclosing. Instead of pre-computing the shape of the outline, the material's intrinsic properties are used to produce complex geometric shapes. The formal result thus resembles the material's capacity and requires a feedback loop between material and structural tests in physical space in the physical model, and a translation into both reliable and controllable parameters in the digital model.

5 Case Studies: Integration in Architectural Concepts

5.1 Connecting Algorithmic Design and Robotic Fabrication

A second generation of test cases takes over previously designed fabrication concepts and re-evaluates these processes in correlation with algorithmic design strategies. A predefined programming and robotic control setup in Grasshopper was combined with the physical toolset of the integrated gripper and feeder system

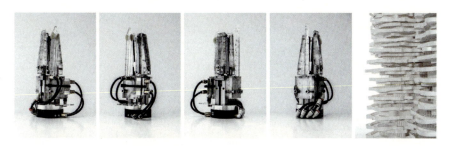

Fig. 13 Tower made out of paper stripes and stapling process, which uses paper's intrinsic properties in the material deformation process (Gramazio and Kohler, FCL Singapore, 2012, student project by Ernst S, Rickhoff S and Strohbach S)

for prefabricated laser cut sheets. The initial setup enables students to work with an unlimited number of different parts from the beginning on. This shifts the focus away from fabrication experiments, and allows them to identify specific typological elements that could be directly translated into parametric design models and utilized to produce differentiated high-rise assemblies.

The students develop computational design engines that are connected to the robotic fabrication process and can start to build multiple iterations of tower models. These models are analyzed for structural performance, material behavior and architectural performance. They are used as the starting point for the next generation of models, thus forming a feedback loop between computational design and robotic fabrication. The resulting designs incorporate several iterations of structural modifications in accordance with a suitable material system. Structural optimization becomes part of the development (Reiser and Umemoto 2006), although this was not the main focus. The following examples show, how different fabrication concepts generate elements that define spaces and express different formal languages.

5.2 Assembly of Folded Elements

This typological concept inverts the traditional approach of conceiving high rises as envelopes and then slicing them into floors by starting the tower design from its interior spaces. Interior spaces are defined by a set of simple rules. The number of shear walls gets increased while their distances decrease. Walls are framing functions rather then rooms, and spaces were defined by cut outs in a sequence of walls instead of being defined by enclosure.

The design code distributes these opening sequences in the walls, which correlate with different apartment and mixed-residential types. In a second step the force flow is calculated from top to bottom around these apertures. Each shear wall adjusts its opening's geometry, negotiating between the calculated force flow and the necessary cut out for the flats.

Fig. 14 The tower is made out of paper elements that are manually cut and then robotically folded, glued and placed on the model (Gramazio and Kohler, FCL Singapore, 2013, student project by Jenny D and Stadelmann JM)

The final tower consists of 5,200 paper elements, which are robotically cut, folded and assembled to a 3 m high tower. The robotic setup contains a mechanical gripper that is able to pick thin paper stripes, a gluing station and a custom designed vacuum clamp for folding. The robot picks a paper stripe and places it into a vacuum clamp at a specific angle, which is computed by the design engine. The students then cut the piece manually with a knife. After that the robot folds the cut piece, applies glue and places the wall onto the model (Fig. 14).

Since the wall elements can be fabricated with cuts in different angles and with individual dimensions, the system is capable of producing a highly differentiated assembly with a single system: on the local scale this leads to different sized spaces and room sequences. In the global system a coherent structural system is achieved while the undulating slab typologies can adjust to contextual parameters. The thin paper adds intriguing light qualities to the model through the large range of variation and the gradual transitions from one element to the next.

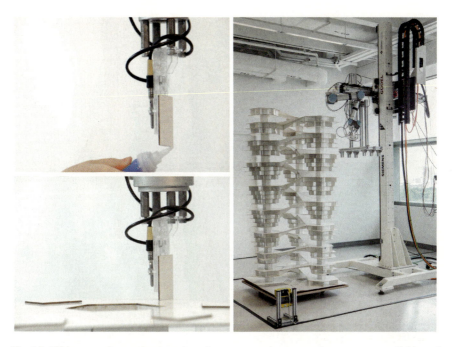

Fig. 15 With sensor integration into the gripper system the process becomes more reliable and faster; the pieces are 'handed' over to the robot that assembles them on the model (Gramazio and Kohler, FCL Singapore, 2013, student project by Foong KQ and Kan L)

5.3 Sensor Integrated Assembly

Sensor assisted picking and placing in combination with robotically controlled thermal bending of acrylic façade elements is used to fabricate this concept for a bundled tower. All data required for the robotic process is generated in the same software environment as the design. The fabrication part itself is not fully automated, since elements are 'handed' over to the robot's gripper. With the integrated sensors the placing of the robotic assembly process allows for tolerances in the material feeding process. Only the centerline of the element must be aligned with the gripper, the sensor measures the vertical distance when placing the element and pushes it just enough to ensure a sufficient contact before the glue hardens (Fig. 15).

For the more intricate, single- and double-bent façade elements a similar deformation process is used as previously described (compare with Fig. 10). For the precise location of these elements, the robot guides the correct picking location by referencing it on the model. Although this fabrication process becomes in large part manual, the assembly is quicker and more robust then the heavily engineered and fully automated process with the feeder for laser cut sheets (compare with Fig. 9). This aspect creates a very interesting dichotomy into fully automated

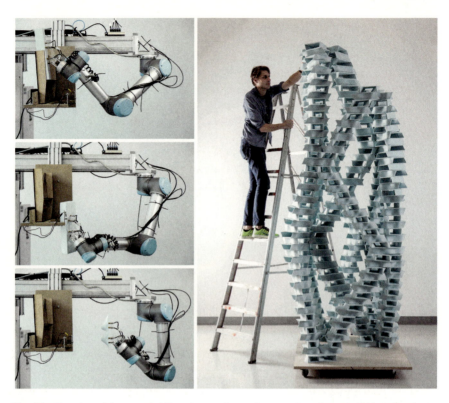

Fig. 16 The robot picks extruded Styrofoam cubes and moves them through the hotwire along a computed path, thus utilizing the robot's potential for performing spatially unrestricted tasks (Gramazio and Kohler, FCL Singapore, 2013, student project by Aejmelaeus-Lindstroem P, Chiang PH and Lee PF)

processes that demand high precision and high amount of debugging, and a mixed manual and robotic processes that benefits from the human interaction for immediate supervision and correction.

5.4 Spatial Wire Cutting

This tower concept consists of a porous mesh with slender tower strands. Several towers merge and separate as they grow in height, structurally supporting each other. Individual residential units are generated on top of each other and oriented away from their immediate neighbors to provide privacy. These units are based on wall elements with complex geometries. Each of them has different geometries, optimised to transfer the loads through the walls and shaped to accommodate various functions on the inside.

As a matter of course the cardboard material becomes limiting for the intended geometries as it proves to be too rigid to be spatially deformed. After several evolutions of gradually more complex articulated wall systems extruded Styrofoam is considered as the solution with most potential to produce the volumetric characteristics. For the final tower, pre-produced cubes are placed on a custom designed feeder. The robot picks the cubes and moves them through a hot wire along a computationally generated path to fabricate the complex wall elements (Fig. 16). As such this process in particular takes advantage of the robot's capability to accurately move and orient in space. The resulting tower consists out of six strands, in each of them up to fifty units with four individually cut Styrofoam pieces. About 2 weeks is the fabrication time for the final tower model for a group of three students.

6 Conclusion

In total twenty-seven 1:50 models were produced in the studios 2012 and 2013. The unique experimental setup gives the architectural model a new meaning and re-values its importance in the combination with digital design tools. The systematic developments of the designs in correlation with the physical artifact obliges students to deeply and creatively engage with the fabrication logic, which becomes a crucial part of the design (Gramazio and Kohler 2008). The construction of iterations of tower models involves continuous feedbacks between physical result and digital design concept. Beyond that models reach a complexity, which could not be manually achieved. The robot thus enables new design explorations in architectural models and it is accounting for the basic condition to avoid a conventional design sequencing, where the data in the end gets handed over to a completely separated fabrication process.

The diversity of the produced towers exemplifies how, after acquiring the skills, computation and digital fabrication can be productively integrated in the design process. The consistent interaction with the robotic process leads to a direct and sensual understanding of the tectonic qualities in the model (Willmann et al. 2012). This exposure to the process of making also requires a profound understanding of the tools and their effects on material and geometric shapes. The role of the architect is challenged here, where design opportunities become sustained in physical space through the adaption and even invention of suitable toolsets. The tested methodology proves to be a valuable experiment on the way to further implementation of robotic fabrication in the design process of large-scale architectural typologies.

Acknowledgments This work is part of a larger research project at FCL and demanded support from Prof Fabio Gramazio and Prof Matthias Kohler, senior researchers Jan Willmann, Silke Langenberg, and co-researchers Norman Hack and Selen Ercan. It was established at the Singapore-ETH Centre for Global Environmental Sustainability (SEC), co-funded by the Singapore National Research Foundation (NRF) and ETH Zurich.

We would like to thank our students for their great efforts, whose projects are illustrated here: Petrus Aejmelaeus-Lindstroem, Pun Hon Chiang, Sebastian Ernst, Kai Qui Foong, Yuhang He, Pascal Genhart, David Jenny, Patrick Goldener, Lijing Kan, Sylvius Kramer, Ping Fuan Lee, Sven Rickhoff, Jean-Marc Stadelmann, Silvan Strohbach, Michael Stünzi, Martin Tessarz, Florence Thonney, Alvaro Valcarce, Fabienne Waldburger, Andre Wong and Tobias Wullschleger. The studio 2013 was conducted under both the ETH Zurich and the National University of Singapore (NUS) curricula. Special thanks go to our academic partners at NUS, Chye Kiang Heng, Yunn Chii Wong, Shinya Okuda and Patrick Janssen.

References

Brayer M-A (2013) Flight assembled architecture: Gramazio & Kohler and Raffaello D'Andrea. HYX, Orléans

Budig M, Lim J, Petrovic R (2014) Integrating robotic fabrication in the design process in architectural design, May/June 2014 No 229. Wiley and Sons, London, pp 22–41

Cousineau L, Miura N (1998) Construction robots: the search for new building technology in Japan. ASCE Publications, VA

Gramazio F, Kohler M (2008) Digital materiality in architecture. Lars Müller Publishers, Baden

Gramazio F, Kohler M (eds) (2014) Made by robots: challenging architecture at a larger scale. Architectural Design, May/June 2014 No 229. Wiley and Sons, London

Hack N, Lauer W, Langenberg S, Gramazio F, Kohler M (2013) Overcoming repetition: robotic fabrication processes at large scale. Int J Architect Comput (IJAC) 11(3):286–299

Kripper R, Barthel R, Petzold F (1989) Wendepunkte Im Bauen: von der seriellen zur digitalen architektur (German Edition). Dresden, Germany

Lim J, Gramazio F, Kohler M (2013) A software environment for designing through robotic fabrication. In: Proceedings of the CAADRIA conference, National University of Singapore, pp 45–54

Menges A, Ahlquist S (2011) Computational design thinking: computation design thinking (AD Reader), 1st edn. Wiley, London

Monkman G, Hesse S, Steinmann R, Schunk H (2007) Robot Grippers. Wiley-VCH Verlag GmbH & Co. KGaA, Darmstadt

Reiser J, Umemoto N (2006) Atlas of novel tectonics. Princeton Architectural Press, New York

Sheil B (2005) Transgression from drawing to making. Architectural Research Quarterly, vol 9, no, 1. Cambridge University Press, Cambridge, pp 20–32

Willmann J, Gramazio F, Kohler M (2014) The robotic touch – How robots change architecture. Park Books, Zurich

Willmann J, Gramazio F, Kohler M, Langenberg S (2012) Digital by material. In: Brell-Cokcan S, Braumann J (eds), ROBIARCH Robotic fabrication in architecture, art, and design. Springer-Verlag, Vienna, pp 12–27

FreeFab

Development of a Construction-Scale Robotic Formwork 3D Printer

James B. Gardiner and Steven R. Janssen

Abstract Despite much recent hype around construction-scale 3D printing, these techniques have been slow to deliver commercial capability even within niche markets. This is in contrast to 3D printing technologies, which are finding applications in the aerospace and medical industries all the way through to jewellery and figurines. Despite the slow start for construction 3D printing there is however significant potential for these techniques within construction industry. Inquisitive and experimental designers are increasingly using sophisticated digital tools to develop complex 3D buildings and sculptural object designs, pushing the boundaries of what can be built. This is resulting in significant challenges for builders to fabricate and build these designs efficiently, especially with concrete. There is also a growing ethical and commercial requirement to create more with less, to use materials only where they are needed, to enable material recycling and to embed objects with greater capability. The FreeFab project is focused on shifting this paradigm with the development of a fabrication technique that produces energy efficient, low waste, fast and cost effective formwork. This chapter describes the development and testing of a novel robotically controlled wax 3D printing system and the software that drives it.

Keywords 3D printing · Robotic · Parametric · Precast concrete · Construction

J. B. Gardiner (✉) · S. R. Janssen
Laing O'Rourke Australia Construction Pty Ltd, Sydney, Australia
e-mail: JGardiner@laingorourke.com.au

S. R. Janssen
e-mail: SJanssen@laingorourke.com.au

W. McGee and M. Ponce de Leon (eds.), *Robotic Fabrication in Architecture, Art and Design 2014*, DOI: 10.1007/978-3-319-04663-1_9,
© Springer International Publishing Switzerland 2014

1 Introduction to Construction 3D Printing

Shortly after 3D printing was invented in the late 1980s researchers realised that the technology could be scaled up to build construction elements and buildings. Construction 3D Printing, also known as 'Additive Manufacturing for Construction' is an emerging construction technique based on methods similar to 3D printing: namely fabricating objects sequentially in layers, while using very different materials, such as concrete. The task of 'scaling up' these technologies, which started in 1996 (Khoshnevis 1996), has been a slow process and one that still has a multitude of challenges ahead (Gardiner 2011). There is an opportunity to leverage the capabilities of Construction 3D Printing now while we wait for the existing techniques to mature or for new ones to appear.

Creating bespoke three-dimensionally curved and/or highly complex precast concrete panels and construction elements today is a costly, time consuming, and a waste generating exercise. This research and development project focuses on leveraging the capabilities of 3D printing for the creation of complex wax formwork without the waste and costs associated with conventional techniques. To achieve this end a novel robotic 3D printing formwork fabrication technique has been developed based on a large 6-axis robotic arm.

The authors draw on extensive primary background research and field testing with construction 3D printing techniques and describe the requirements that underpin this project within the realm of construction 3D printing and a commercial construction enterprise. The key constraints that influenced the development of the 3D printing technique are described along with findings both positive and negative from the short period of prototyping since assembly of the system.

2 Bespoke Precast Concrete

Creating formwork for concrete is a relatively time consuming exercise and the more complex the cast, the more time is required to build the mould. Contemporary casting methods for freeform and complex bespoke moulds include the use of milled polystyrene or cast polyurethane/silicon moulds. Polystyrene moulds, used for one-off castings, are generally milled using a CNC router or a robot. After being removed from the cast concrete, the polyurethane mould is sent to landfill. Cast polyurethane and silicone moulds, largely used for serial production, are expensive and also require the fabrication of a secondary mould for casting the silicon or polyurethane moulds into, generating more waste that ends up in landfill.

There are a number of groups that have looked at alternative casting techniques to use for creating freeform and decorative concrete panels, such as fabric formwork by Mark West of C.A.S.T (Beorkrem 2013) and Gamazio and Kohler of the ETH Tailorcrete project. Each of these processes use cheap and recyclable base materials to create the bespoke formwork for creating concrete panels, however

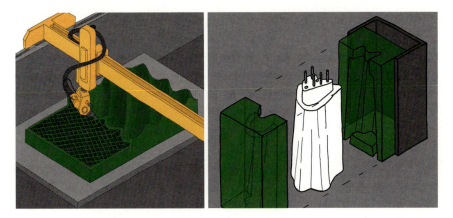

Fig. 1 Wax formwork 3D printer fabricating a hollow mould for a panel with complex surface geometry (*left*); a 3D printed wax mould used as a sleeve for a free-form column casting (*right*)

neither offer a process that is broadly applicable to producing a wide variety of moulds, from the intricate detail of the Noppenhalle by Baierbischofberger Architects to the large-scale form of the Spenser Dock bridge by Amanda Levete Architects.

3 FreeFab Wax 3D Printing Technique

The concept for the wax printer is to create a process that uses as little energy as possible to create an infinite variety of high quality freeform wax moulds that are completely recyclable. In order to be commercially viable, the process must be fast and also highly accurate (in construction terms). During early testing it was clear that wax is very difficult to melt in block form and holds a significant amount of latent heat after printing due to the phase-change nature of the material. In order to minimise energy use, the concept was developed to create hollow moulds (Fig. 1) that allow dispersion of heated air or water internally to warm them during fabrication and to melt them once the concrete has cured. Internal support geometry is also required to give the moulds a high bearing strength while using as little material as possible to minimise printing time and reduce the energy required to melt the material off for recycling.

Speed is essential to fabricate large moulds (up to 6 m × 4 m) economically and thus rate of material deposition will be favoured over surface quality with a milling spindle being used to finish the mould surfaces as necessary. The heat of the curing concrete is also being considered to melt off the wax moulds, although the timing for this would need to be tightly controlled. The wax will most likely be de-moulded from the concrete within a hot water bath, allowing for concrete dust and other foreign matter to be separated from the wax, which floats. Further filtration of the wax would be completed before re-use.

4 Software Interface

4.1 Robot Control for Wax 3D Printing and Milling

The interface between the design and the robot and its tools is a key element for the creation of a viable and integrated fabrication system. In order to control a robot to complete the 3D printing and milling tasks discussed above, a review of existing methods for generating robot control code and the associated software packages was conducted. What we were looking for was a seamless integration between the CAD software and the robot, to reduce the inefficiencies caused by passing information between multiple software packages and invariable compatibility issues.

In research conducted to determine the most efficient interface between CAD software and the robotic hardware the chapter titled *A New Parametric Design Tool for Robot Milling* (Brell-Cokcan and Braumann 2010) was particularly valuable. Brell-Cokcan describes the traditional process for fabricating parts using an articulated robot arm with a milling attachment (Fig. 2). This process involves creating a digital model of the designed object using a CAD application; generating tool paths over this model using CAM software; converting these tool paths into robot code using a Postprocessor; simulating this code to validate the robot movements; and finally uploading the code to the robot controller, which instructs the robot to carry out the task. This process typically involves iterating back and forth between several discreet software packages until all geometry, reachability, collision and singularity problems have been ironed out.

Brell-Cokcan proposed a new approach to robot control, where the tool paths, robot code and simulations are all generated directly within existing CAD software. This approach simplifies the process of transferring data from CAD software to the robot by eliminating the need to push data between many software packages.

Two software tools, which have since been developed, achieve this workflow for a CAD package called Rhinoceros™. Kuka prc™ can be used to generate code for Kuka robots and HAL™ is used for ABB robots. Both are built into Rhino's powerful parametric scripting language, Grasshopper™. They are conceived as toolkits of additional components, which extend Grasshopper's capability to generate tool paths for milling strategies, output robot code, and detect collisions, reachability issues and singularities.

The advantages of building these new robot control tools into a scripting language such as Grasshopper™ are that they allow the code to be easily connected to existing parametric geometry, and the toolkit can be easily expanded using custom scripted components to achieve any additional functionality that is required.

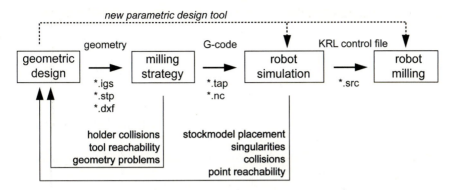

Fig. 2 Traditional robot control pipeline as described by Brell-Cokcan and Braumann (2010)

4.2 Creation of the Digital Model for Wax 3D Printing and Milling

Using the HAL^{TM} software tool described above for ABB robots, we developed the following process for generating robot code for wax 3D printing and milling (Fig. 3). Three distinct methods for creating digital models for robotic 3D printing and milling were investigated. The first method involved hand-sculpting a block of wax to the desired form and using a laser-scanner to digitise the model. This was then cleaned up in Rhino^{TM} before being linked to the Grasshopper^{TM} script. This method would be most appropriate for restoration and heritage construction work.

The second method involved modelling the desired form using T-splines for Rhino^{TM}. This software allows for completely freeform design of building elements including columns, walls and beams. The model can be linked to the Grasshopper^{TM} script, allowing changes to be made while the robot code and simulation update in real-time. This method is valuable when the designer requires complete artistic control over the final form, but wishes to stay within the limitations of the robotic system.

The third method for creating digital models was to generate the model parametrically within Grasshopper^{TM}. This method was found to be particularly beneficial for developing surface patterning and textures, when the designer needs to iterate quickly through many variations to find the most visually appealing result or multiple iterations and variations are required.

4.3 Generation of Tool Paths for Wax 3D Printing and Milling

Once the digital model has been linked to Grasshopper^{TM}, tool paths can be generated over the model. In the case of 3D printing, the model is sliced into horizontal layers, filled with internal support geometry and broken down into a list

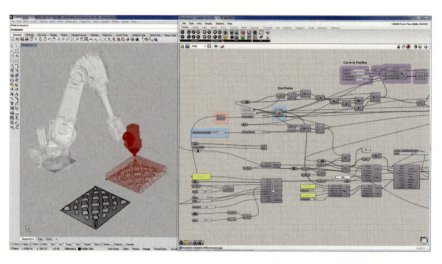

Fig. 3 The software interface for generating and simulating robot control code for the wax formwork 3D printer

of targets for the robot to reach. The outline of each layer is taken and offset multiple times to generate the required external wall thickness, based on the width of each bead of wax. In our initial experiments, the bead was approximately 4 mm thick and as an external thickness of 20 mm was desired, five layers were required to build up this thickness. Starting and stopping the flow of wax resulting in a small blob of extra material at the start and end of each tool path, so the five exterior layers were connected into a continuous tool path (Fig. 4) to minimise these blobs. This geometry was achieved using standard components within Grasshopper™.

Support geometry was then generated to fill the internal void of the object. Again, numerous layers were used to obtain the desired thickness. Initially, a square grid pattern was used, but it was found that sudden changes in direction introduce vibrations into the extruding head and robot, so the pattern was modified into a series of sine waves that the robot can follow more smoothly (Fig. 5).

External and internal tool paths were then broken down into discreet points, through which the robot would move. The spacing between these points was optimised based on the curvature of the path (Fig. 6). As the accuracy of the robot is usually not better than 0.05 mm, points were also not spaced closer than this.

4.4 Conversion of Tool Paths to Robot Code

The HAL™ components within Grasshopper were used to convert target points into robot code. After being verified through simulation, this code could be uploaded to the robot controller. A limitation of the controller is that it can only

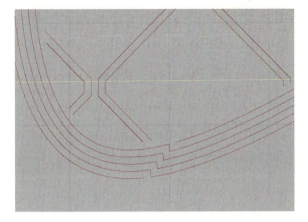

Fig. 4 Detail of tool paths for the exterior wall

Fig. 5 Square grid internal geometry patter (*left*); wobble visible in the 3D print due to vibrations in the wax extruder head (*centre*); sine wave internal geometry pattern (*right*)

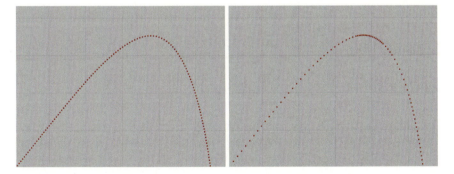

Fig. 6 Even spacing of targets along the tool path (*left*); optimised targets along the tool path (*right*)

load code with less than about 60,000 lines of code, more than this the robot throws an error. This limit can be easily exceeded during a 3D printing and/or milling application. To overcome this obstacle, a custom component was developed to split the code generated by HAL^{TM} into multiple code modules, each of which contained less than 60,000 lines. These modules are then sequentially loaded and unloaded by a master module on the robot controller. The code must be carefully separated so that the robot loads each new module at the end of a tool path, when the extrusion nozzle is turned off. A new technique is also being investigated to load/unload the modules in the background to avoid pauses during printing while files are loaded into the controller.

4.5 Simulation of Robot Code

The robot code is used to simulate and validate the robot movements. The simulation built into HAL^{TM} was used to visualise the robot at each of the targets and check for 'target out of reach', 'joint out of range' and singularity issues as well as self-collisions and collisions with surrounding objects. This provided an early opportunity to correct these issues before uploading anything to the actual robot and without the need for passing data between different software packages.

While HAL^{TM} can also be used to predict abnormal reorientations; this was found to return many false positives. The reason for this is because HAL calculates each position statically, ignoring the targets before and after the current one. By contrast, the actual robot controller calculates the next position dynamically based on the current position. For a more rigorous simulation of these effects and for visualisation of the robot accelerating and decelerating between each of the targets, the robot code generated by HAL^{TM} was loaded into ABB's Robot Studio software and tested. Often it was found to be easier to load the code into the robot controller and run the program at reduced speed, with no work piece and with the extrusion head and milling spindle disabled, hence negating the need to introduce another piece of software into the process.

5 Prototyping

5.1 Equipment

The prototype cell developed to test the viability of the system is significantly smaller than the scale of the commercial system, although the hardware (such as the robot and the wax melting system) will likely be within the same class and sourced from the same suppliers. The production scale system is intended to fabricate wax moulds 6 m × 4 m along with other large complex moulds for non-planar elements. This scale of production would require the use of multiple hanging, track-mounted robots above an automated carousel.

The system developed for testing the commercial feasibility of the system is built around a 6-axis ABB 4600 robot with a 2.05 m reach and a 60 kg payload. This robotic arm has been integrated with both a high speed milling spindle (with an automatic tool changer) and a wax printing nozzle. These two primary tool attachments can be automatically interchanged during a build operation to enable tandem wax printing and finishing during the build cycle, without manual intervention. The prototype system allows for moulds up to 1.2 m × 1.2 m to be created.

5.2 Materials

Wax was selected as the build material for the system. This material was chosen from a wide variety of potential materials for a number of reasons: wax is a very versatile material that can be adjusted to exhibit a range of material properties as required. Some advantages of using wax include it being:

- well suited to casting
- easily recyclable
- comparatively inexpensive
- able to be worked/milled to achieve a high quality finish.

Wax materials have been used for 3D printing at small scale for many years, especially for creating jewellery using lost wax casting. Wax has also been used for large scale concrete formwork (Maeda 1999). The two have not, however, been brought together previously; it is with this and further process innovations that the novelty of this project lies.

The most significant hurdle to date with the development of the wax printer system has been in the development of the wax formula, as off the shelf waxes have not performed well. There are a number of competing requirements for the wax, such as

- low melt temperature (to reduce energy consumption)
- high viscosity (to allow printing without support materials)
- low shrinkage (to ensure dimensional accuracy of the wax formwork)
- Machinability (to allow milling to remove imperfections in the 3D printed wax).

Developing and optimising the wax formula to meet these requirements is requiring significant efforts in-house and from our wax supplier to resolve; solving one problem often exacerbates another and thus a series of different formulas have been tested as we push toward an optimal solution. Delamination between deposited layers is a common issue with smaller scale fused deposition 3D printing systems and at larger scales this issue and others are magnified.

Recyclability of the wax is another important consideration for the development of the wax 3D printing system. Currently, several methods are being investigated for the retrieval of the wax from the concrete panel. Early experiments recovered up to 93 % of the melted wax for recycling.

5.3 Testing

The majority of testing over the last 3 months since the setup of the system has focused on milling, which will form an integral part of the finishing processes. The milling component has been largely successful, with code generated directly from the HAL[TM] plugin for Grasshopper[TM]. The only major obstacle encountered with milling to date related to the robot memory limit (an issue common to both milling and 3D printing), which was mentioned above in the discussion on digital process.

During testing of the 3D printing process several issues were encountered. Figure 7 shows that if the viscosity of the wax is too low, the wax drips down the sides of the print whilst printing subtle overhangs. This would require much cleaning up with the milling spindle before the mould could be used. To compensate for this, the temperature can be lowered, resulting in a more viscous wax. This, however, tends to lead to delamination issues between the printed layers. Finding a viscosity and temperature combination that produces properly bonded layers without too much dripping is a delicate process.

Another issue encountered during the 3D printing process was warping that caused a 1,200 mm wide 3D printed object to delaminate from the build platform (Fig. 8). This issue could be solved in a number of ways, such as using a heated platform or applying adhesive to the base prior to printing. The problem with this build however was more pronounced than in others due to the addition of a new material to the wax formula to increase viscosity, hence solving one issue often causes others. It can be illustrated with the two examples above that the solution to one issue invariably has an effect on another and hence a process of optimisation is occurring whilst looking to new formulas that have minimal impacts on desired requirements.

Figure 9 shows a 3D print with multiple outer layers and an internal support geometry that is changing colour as it cools in temperature. The outer surface is built first with a continuous stepped bead starting on the inside and finishing on the outside to reduce the effect of splutter when the nozzle opens. The outer layer is the casting face and is designed to be both strong and have an allowance for milling to achieve a high quality surface finish.

Fig. 7 Wax print showing dripping caused by using low-viscosity wax

Fig. 8 Shrinkage causing warping of the wax object from the build platform of large multilayer print

A benefit of the integration of the wax printing process with a milling spindle is the ability to print very quickly, sacrificing some quality of surface finish for an increase in material deposition speed, which is considered vital for fabricating very large moulds for concrete panels. The milling spindle can then be used to quickly tidy up the casting face of the mould, which in most cases will be a small fraction of the total surface area of the mould. A second benefit of having an integrated milling head within the process is the ability to remove cumulative geometric distortions during the build (Fig. 10). At the intersection points in the print, slightly more wax accumulates than elsewhere; over 20 or more layers this effect is clearly visible and if not corrected would cause significant distortion wax mould. The milling spindle is used after a set number of layers to flatten the top surface of the 3D print before continuing with the rest of the print.

Fig. 9 First layer of 3D printed wax build (*left*); an earlier build after approximately 50 layers (*right*)

Fig. 10 Cumulative distortion in the 3D print

6 Conclusion

The development of this novel large scale 3D printing technique for the creation of complex wax formwork has the potential to unlock many of the capabilities of construction 3D printing, while eliminating a number of issues that have slowed their development to date, especially relating to structural capability in finished elements.

Within this chapter a novel construction scale wax printer has been described that has the capability to fabricate large wax moulds for creating precast concrete elements. The system has been designed to minimise waste and energy use whilst delivering capability for fabricating formwork for a wide variety of precast applications. The prototype system has demonstrated strong potential while revealing several of the challenges that lay ahead, especially in relation to the wax formula, where competing requirements are proving a challenge to balance.

The project has implemented the use of the HAL^{TM} software component for Grasshopper^{TM} that allows a relatively direct interface between the Rhinoceros^{TM} design software and the ABB robotic system. While the HAL^{TM} software has proven to be effective in enabling this process, significant custom scripting has been required to enable effective use of this interface and perform the tasks required.

The development of the patented process described is a rare example of a large construction company making significant direct investment into innovation at the frontiers of research, within an industry with low R&D expenditure relative to other industries (Manseau and Seaden 2001).

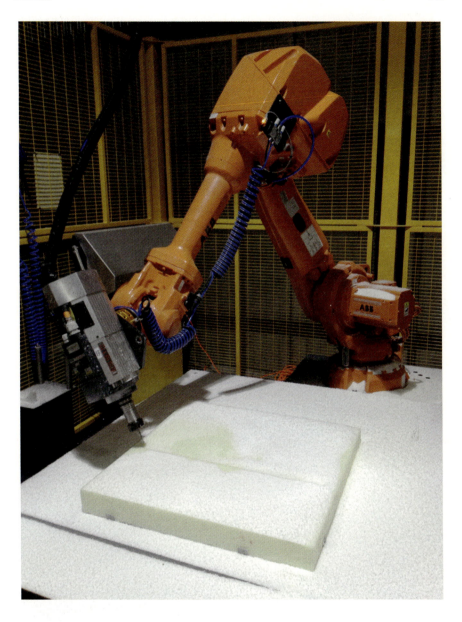

References

Beorkrem C (2013) Material strategies in digital fabrication. Routledge, NY

Brell-Cokcan S, Braumann J (2010) A new parametric design tool for robot milling. In: Paper presented at the 30th annual conference of the association for computer aided design in architecture (ACADIA), New York, pp 357–363

Gardiner JB (2011) Exploring the emerging design territory of construction 3D printing—project led architectural research In: PhD thesis, RMIT University, Melbourne, pp 362–364

Khoshnevis B (1996) Additive fabrication apparatus and method. US Patent 08/382,869

Maeda N (1999) Method for manufacturing concrete product. Japanese Patent 2001-105420

Manseau A, Seaden G (ed) (2001) Innovation in construction: an international review of public policies. Taylor & Francis Books Ltd, New York

Additive Manufacturing of Metallic Alloys

An Architectural Case

James Warton, Rajeev Dwivedi and Radovan Kovacevic

Abstract Additive processes have gained increasing interest within the discussion of digital fabrication and architecture. In general, architects have encouraged limited implementation of this relatively new mode of production beyond conceptual and representational applications. Few examples exist that pursue the large scale application of this technology and these few cases are primarily focused on utilizing polymer and resin-based materials, ceramics, sand and cementitious materials. While additive manufacturing of metals has reached production scaled efficiency and cost feasibility within medical, aerospace, and aviation manufacturing industries, it has yet to make a significant presence as part of architectural discourse. This chapter presents two additive methods of manufacturing in metals most commonly employed for production and their possible applications within the architectural project. The research presented explores the expanded territory for design freedoms, as well as the higher degree of optimizations unique to this manufacturing process. In addition target areas are defined for implementation within a fully integrated design to manufacturing solution space.

Keywords Additive manufacturing · Metals · Functionally graded material · Lattice structures · Finite elements method

J. Warton (✉) · R. Kovacevic
Southern Methodist University, University Park, TX, USA
e-mail: jwarton@smu.edu

R. Kovacevic
e-mail: kovacevi@lyle.smu.edu

R. Dwivedi
Finisar Corporation, Allen, TX, USA
e-mail: rajeev.dwivedi@finisar.com

W. McGee and M. Ponce de Leon (eds.), *Robotic Fabrication in Architecture, Art and Design 2014*, DOI: 10.1007/978-3-319-04663-1_10,
© Springer International Publishing Switzerland 2014

1 Additive Manufacturing in Architecture

Presently several approaches to additive fabrication are being explored for architectural applications. D-Shape (Dini et al. 2006), one of the most notable large scale projects, employs a mineral powder composite and uses an inorganic compound to bind layer upon layer similar to the process found in smaller selective laser sintering (SLS) printers. Other areas of research focus on a process similar to fused deposition modeling (FDM). IMCRC (Kestelier 2012) and USC (Khoshnevis et al. 2006) employ a model closely following this process and utilize a semi-solid cementitious material rather than polymer filament. More recently IAAC introduced (http://www.mataerial.com) research using a free-form process that breaks with the more common stratified layering processes. This process is closely aligned with a free-form path planning approach and direct metal deposition (DMD) processes implemented by Trumpf (http://www.youtube.com/watch?v=iLndYWw5_y8) and Lockheed Martin for aviation and aerospace parts production.

Like other forms of rapid prototyping and additive manufacturing, those utilizing metals offer compelling potential for designers. Among these are the integration of complex and intricate surface textures, variable material density, and heterogeneity of material substances within a single build (Oxman 2010). These additive manufacturing technologies address a number of functional criteria such as the production of lighter, cheaper, waste reduction alternatives to traditionally manufactured parts (http://www.theengineer.co.uk/in-depth/the-big-story/the-rise-of-additive-manufacturing/1002560.article). The capacity to print metal parts with variations in material properties as well as a combination of heterogeneous material composition have also allowed for single parts to replace whole assemblies (Rockstroh et al. 2013). Because additive manufacturing is so well suited for producing parts that demand strength with minimal weight, it has gained increasing attention in aviation, aerospace, and medical industries (Wholers 2011).

At this juncture it is necessary to differentiate this research agenda from that of previously mentioned contemporaries. This exploration is not simply a divergence in material or production oriented processes as related to implementation of 'free-form construction'. A more relevant distinction stems from how these factors enable a high resolution architecture of structural intricacy and lightness unrestrained by a material logic embedded in compression loading sensibilities. One of the key aims behind this research is the development of visually light, seemingly transparent and highly articulated structural assemblies enabled through the interrogation of a novel approach to manufacturing.

Inherent in the foundations of architecture theory are explorations of the relationship between material properties, methods of construction and the ideal forms they encompass. Nineteenth century French architect and theorist Viollet-le-Duc was among the first to articulate the idea that the development of new materials and material processes demand proactive engagement to further the production of architectural form:

We shall now attempt to enter more fully upon the employment of novel materials, and to deduce therefrom certain general forms of construction under novel conditions...

...making use of materials in accordance with their respective properties; with a frank and cordial adoption of industrial appliances, and instead of waiting for these to take the initiative, ourselves eliciting their production (Hearn 1990).

His assertions serve as a relevant position for the exploration of an architecture enabled through untested approaches to material, and manufacturing. It also implies a clear distinction from those cited in regards to deeper formal implications. Validation of this hypothesis begins with a closer look at available additive methods using metals and extends into speculations of how these methods can be appropriated and customized to serve the discipline of architecture.

2 Additive Methods for Metal

There are fundamentally two approaches governing the mechanics of most additive manufacturing systems which utilize metals. The most common are the 3-axis systems utilizing partially sintered material as its primary support medium. These machines produce very high resolution parts with a tolerance of 0.002 mm (Fig. 1), minimal material waste, and are effective for functional prototypes, design feature testing and validation, as well as end product fabrication. An Arcam A2 machine was employed for these 1:1 prototypes comprised of titanium (Ti-6Ai4v) (Fig. 2). This machine utilizes an electron beam melting (EBM) process to fuse powder into fully sintered part geometry. Due to the chamber size (200 mm × 200 mm × 300 mm), parts produced on this machine are limited in scale. Despite this limitation, EBM produces parts ideal for use within non-standardized assemblies such as cable-net structures, point supported glazing systems, and nodes within various forms of lattice structure and space frames.

Another effective approach employs the use of numerically controlled robots, high powered fiber-optic lasers, and material feeding systems. Material is deposited directly along the build path and fused at its point of contact with a laser beam. This type of system may deposit powder and/or wire-fed material (Syed et al. 2006). These free-form fabrication systems enable production within a large build area suitable to components within larger span structures. Because of their broad range of motion, path planning need not be limited to parallel layering. Unlike the alternative large scale deposition systems previously mentioned, Free-Form-Path DMD allows for higher tolerances and yields a near net shape part with minimal post processing requirements to achieve a more refined end product.

Metals can be deposited relatively quickly in comparison to cementitious material and do not face the challenges and deformation caused by overloading partially cured material. Newly deposited powder becomes fully fused to the substrate almost instantly, and can be deposited along paths that bridge across unsupported regions. Within certain limits the surface tension and solidification/cooling rate prevent molten areas from drifting outside a narrow distortion

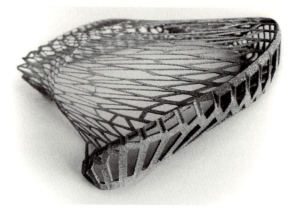

Fig. 1 Procedural lattice: prototype featuring nodes with variable branch count, branch depth and thickness. This hypothetical design for highly articulated structural frame demonstrates effective wire-like branches 1.5 mm thick spanning just over 30 mm

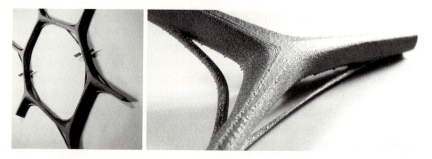

Fig. 2 Titanium prototypes: assembly composed of non-modular components (*left*) and individual *y*-branch component (*right*). Each node features unique configuration

tolerance. Because deposition to the substrate is somewhat directionally independent, build orientation is primarily limited by collision prevention. The ability to build parts with unsupported overhangs that cantilever from previously fused material enables for higher degrees of design complexity, while minimizing waste associated to extraneous support. In addition, deposition rates, multiple feeding systems, laser intensity, and scanning velocity can be calibrated to regulate variable material composition and thickness along its build path.

3 Multi-Scaled Performative Expression

Creating a multi-scaled form of structural articulation is a key area of interest for application of this technology. This combination of materials and fabrication process enables a precise and calibrated engagement with structure at scales

Fig. 3 Hollow *Y*-branch prototype fabricated using electron beam melting (*EBM*), and cross sectioned to reveal internal supports modeled after avian bone features

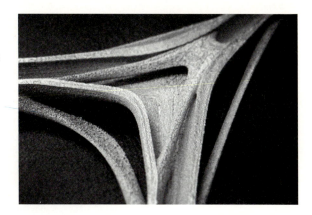

ranging through the micro-level organization of material compositions to the macro-level organization of non-modular assemblies. Opportunity extends beyond the ability to address strict functional optimization and variable loading criteria enabling expressions of complexity and performance synthesis which traverse the full range of architectural scale.

Optimization objectives center on achieving visual lightness through a vocabulary of delicate intricacies and multi-scaled fields of density. Strategies address stress accumulation and buckling resistance within novel or 'arbitrary' free-form design. Whether developed algorithmically or through an iterative modeling process, design features must be employed to reciprocate these performance criteria. One of the fundamental advantages of additive manufacturing is the ability to define internal complexities that address a range of performance needs. An understanding of these potentials will inevitably lead to unique features such as hollow structures with internal stiffening resistance (Fig. 3).

Thin wall optimization based on morphologies present within avian bone structures were examined as a possible design model. This may prove beneficial for light-weight structures with stiffness and buckling resistance unmatched by equivalent cross sections produced through extrusion (Fig. 4). The skeletal structures of birds feature a number of adaptations that have produced an extremely light-weight structure capable of tolerating stress imposed during take-off, flight, and landing. One key adaptation is the fusing of bones into a single ossification allowing for the overall reduction of bone mass. In addition, many bones are hollow with an intricate network of reinforcing struts. These adaptations can be characterized by a variation of bone density consistent with its performance. In many cases these cavities serve a dual function as respiratory air sacs promoting similar functional integration.

Achim Menges (2008) addresses the "high-level functional integration" possible through additive processes in his essay "Manufacturing Performance". It is clear that the impact of an inverse relationship between scale and resolution exist when dealing with construction-scale additive manufacturing. As Menges aptly notes, production times necessitate systems capable of bulk and high resolution

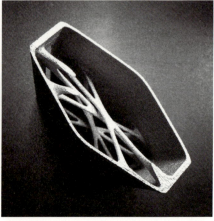

Fig. 4 Thin wall optimization and internal lattice. Cross section through avian bone structure (*left*). Titanium prototype featuring internal supports modeled after avian bone features (*right*). Principle wall thickness is 1.5 mm

deposition rates in order to effectively deliver highly articulated and functionally integrated components at a large scale.

As opposed to the compression materials cited within his essay, metal performs in a manner that reduces the range of this inverse relationship. Slender structures that act in both tension and compression demonstrate the 'free-plan' alternative to load bearing mass while allowing for the integration of functions, such as drainage or rain water harvesting, electrical and data conduits, chase-ways, and plenum space within a single structural system. This approach effectively promotes the objective to achieve a transparency of structure by reducing spatial clutter or visual pollution.

The prototypes based on this model of performance developed in conjunction with iterative FEM analysis starting from the solid model of a hypothetical node (Fig. 5). Stress concentrations between both solid and hollow versions were compared. Thickness variations and additional struts were incorporated in response to the observed stress concentration patterns. The initial test case started with two 3,000 N loads and demonstrated a maximum localized stress value of 315.57 MPa. These initial loading criteria were satisfied by the final design iteration with a 35 % reduction in material volume. Displacement was reduced from 3.88 to 1.92 mm while a factor of safety was increased from 2.79 to 2.98. Destructive testing was not performed prior to submission of this chapter.

Another compelling capacity of additive fabrication is the implementation of functionally graded material or FGM (Fig. 6). The metals used for direct metal deposition include, but are not limited to aluminum, steel, and titanium. The micropowders available to this technology provide additional optimization scenarios. Through numerically controlled variable deposition, material properties can be modulated to address various design criteria and offer a high degree of structural

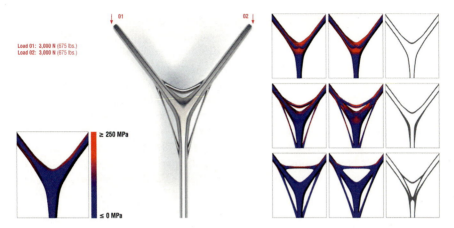

Fig. 5 Stress accumulation observed in solid and uniform wall thickness designs inform subsequent iterations incorporating variable thickness and stiffening members

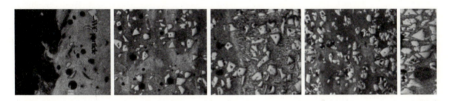

Fig. 6 Cross section showing micro-structural and compositional gradients of a heterogeneous alloy comprised of tool steel and WC-(NiSiB) ceremet powder achieved through variable deposition rates within a single DMD build. (Kovacevic et al. 2002)

performance. Stiffness, hardness, fire and heat resistance and corrosion resistance can be addressed comprehensively through FGM. Base metals can be selected for minimized cost while a combination of other elements such as Nickel, Chromium, or Tungsten Carbide can be supplemented directly where needed in the form of powders producing structural components comprised of several alloys in a single direct to part fabrication process (Kovacevic et al. 2002).

In order for these opportunities to be actualized; path planning and material deposition processes must be developed further. Current CAD/CAM applications have proven insufficient for realizing multi-scaled arrangements of spatial voids and heterogeneous material distribution (Payne and Michalatos 2013). Likewise, Building Information Modeling (BIM) platforms have not anticipated such a high-performance approach to systems integration. The broad range of modeling software offered relies on a boundary representation or b-rep definition based on the general assumption that each part is comprised of a single homogeneous material. The complexity of features possible through additive manufacturing demands the development of a computational approach that integrates simulation, heterogeneous distribution and parametric control.

Point cloud distributions using computational fluid dynamics, particle attraction, and Turing pattern simulation through reaction diffusion and morphogen gradient models may prove effective for driving material distribution; however, this requires programming efforts that are in their initial stages of development. Turing's reaction–diffusion model has already been widely accepted among biologist for researching cell and tissue growth (Yoshimoto and Kondo 2012) and has proven viable for simulation of bone growth (Zhang et al. 2013).

Because this algorithm is well suited for distributing multiple 'species' within a given domain, it may be well suited for this task. This material distribution is not simply a matter of generating effective aggregation relative to stress accumulation or other functional performance criteria, but rather an expanded creative domain where a range of topological configurations and surface effects may be achieved through visual patterning of alloys with varied textural, tonal, and specular qualities and may be propagated at a range of scales. Work within this area has run parallel to research using additive processes and requires further development prior to implementation, performance testing, and validation.

4 Digital Craft

In Kolarevic's (2008) citation of David Pye on the topic of 'digital craft', he demonstrates key differences between the workmanship of manufacturing defined by 'certainty' and the workmanship of 'risk' associated with craftsmanship. Kolarevic further clarifies that the risks are not merely a set of material consequences, rather the uncertainty in regards to outcome. This distinguishing relationship becomes evident when relatively uncharted processes are adopted. While the tools for making are marked by evident areas of material freedom, there are varied and opposing constraints to all processes.

Within this domain of uncharted constraints a formal aesthetic may develop (Hearn 1990). Similar to the ways the associative relations or constraints are selected within a parametrically defined model, the path planning sequence and tooling act as an associative process driving possible formal outcomes. Consequently, therein lies the need for a well-crafted selection of design tools and hardware.

Path planning for DMD is marked with a number of challenges or 'risks' that are not common to planning strategies developed for the production of milled or machined components. Some of these challenges have been addressed and clearly delineated through the research of Dwivedi and Kovacevic (2006). These challenges become further compounded through the integration of additive and subtractive processes within a single fabrication system (Kovacevic and Valant 2006). Build orientation and collision prevention are key players within this process. Their proposed medial path planning approach has clear benefits related to the production of complex topologies and limits the need for additional support material. In these cases, deposition paths follow the geometries medial path

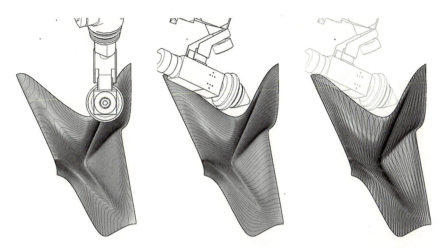

Fig. 7 Path sequence based on surface contours in *xy*-plane and tool center point oriented along *z*-axis (*left*). Path sequence developed from UV parameters and tool center point aligned to surface normal (*center*). Novel path sequence developed from UV parameters on surface geometry (*right*)

alleviating the occurrences of unsupported islands associated to deep undercuts. In addition, when path planning is associated to part geometry either through medial path or surface vectors, tooling and deposition inscriptions reinforce the surface geometry, as opposed to path planning based on external reference systems such as contouring.

Within this strategy, these latent inscriptions inflect the inherent nature of the geometry reaffirming its mathematical description (Fig. 7), whereas the superimpositions of a 'world' x–y plane in some cases create undesirable results. This is characterized through surface quality differences observed in these test cases (Fig. 8). Paths associated to the surface's domain also lead to opportunities for articulated and discontinuous descriptions of surface geometry within a continuous domain while remaining independent from superfluous support material.

Beyond surface rationalization scenarios, branching typologies represent another class of path planning explorations. Small scale implementation have demonstrated this approach is viable for fine features thin as 1 mm (Fig. 9) and an n-branch base class was developed to explore this capacity as a proto-geometry for further path planning explorations (Fig. 10). Additional testing has been implemented through the fabrication of a hollow spring composed of titanium and was built by following a spiraling path (Fig. 11). While these initial tests are small in scale, it's clear that this path planning strategy has potential to yield results effective for the fabrication of intricately detailed structural members.

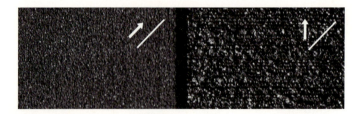

Fig. 8 Notable surface quality improvements achieved when build path is oriented to surface. *Arrow* indicated build orientation relative to part surface

Fig. 9 Branched structure with linear segments fabricated by free-form DMD (Dwivedi and Kovacevic 2007). Deposition oriented to medial path (*right*) and matrix of branch configurations with variable count and thickness (*left*)

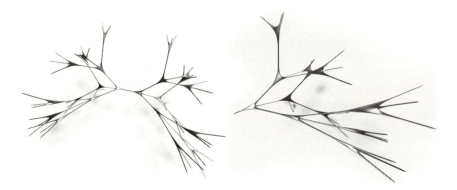

Fig. 10 Tree structures developed from parametric branching node definition

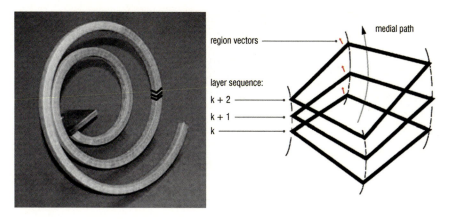

Fig. 11 Hollow spiral structure fabricated through DMD using 6-axis robotic arm using medial path approach (*left*). Enlarged path diagram showing point to point sequence of sections following medial path (*right*). (Dwivedi and Kovacevic 2006)

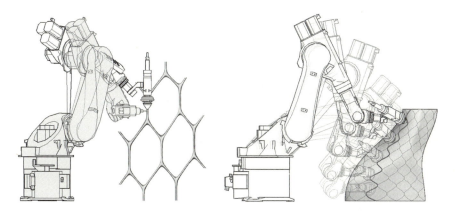

Fig. 12 Additive and subtractive processes coordinated within the same build. Additive oriented along *z*-axis or medial axis and milling oriented to surface normal (*left*). Robotized DMD with deposition orientation following surface normal and medial path (*right*)

5 Conclusion and Future Work

The implementations of additive manufacturing presented clearly yield an expanded domain for design exploration. The range of design tools and software that harness these potentials are limited and integration of a fully functional manufacturing center is in the process. To achieve the stated goals we have initiated the development of an integrated design to fabrication process that employs various feedback mechanisms between FEM analysis and path planning criteria. Surface quality and demands for milled quality finish has led to integrated additive and subtractive path planning (Fig. 12). Architectural applications for additive manufactured metals and free form deposition are broad, and much work is needed to realize an assembly of structural components which comprehensively exemplify many of the features demonstrated here as separate areas of investigation. Through this exploration we anticipate the realization of intricate and high-resolution structural systems that embody a delicate and visual lightness responsive to embedded performance criteria.

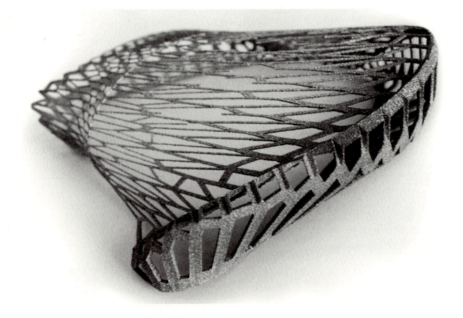

References

Dini E, Nannini R, Chiarugi M (2006) Method and device for building automatically conglomerate structures. WO Patent WO2006100556 A2

Dwivedi R, Kovacevic R (2006) An expert system for generation of machine inputs for laser-based multi-directional metal deposition. Int J Mach Tools Manuf 46:1811–1822

Dwivedi R, Zekovic S, Kovacevic R (2007) A novel approach to fabricate uni-directional and branching slender structures using laser-based direct metal deposition. Int J Mach Tools Manuf 47:1246–1256

Hearn MF (1990) Applying new architectural principles. In: The architectural theory of Viollet-le-Duc: readings and commentary. MIT Press, Cambridge, pp 231–237

Kestelier XD (2012) Design potential for large-scale additive fabrication, free-form construction. In: Fabricate: making digital architecture, 2nd edn. Riverside Architectural Press, Cambridge, pp 244–249

Khoshnevis B, Hwang D, Yao KT, Yeh Z (2006) Mega-scale fabrication by contour crafting. Int J Ind Syst Eng 1:301–320

Kolarevic B (2008) The (risky) craft of digital making. In: Manufacturing material effects, rethinking design and making in architecture, Routledge, New York, pp 42–47

Kovacevic R, Valant M (2006) System and methods for fabrication or repairing a part. Patent US7020539 B1

Kovacevic R, Mei H, Ouyang JH (2002) Rapid prototyping and characterization of a WC-(NiSiB alloy) cermet/tool steel functionally graded material synthesized by laser cladding. In: Rapid prototyping of materials, The Minerals, Metals and Materials Society, Warrendale

Menges A (2008) Manufacturing performance. In: Architectural design, pp 121–128

Oxman N (2010) Structuring materiality: design fabrication of heterogeneous materials. Architectural Des 80(4):78–85

Payne AO, Michalatos PM (2013) Working with multi-scale material distributions. ACADIA 2013 Adaptive architecture, pp 43–50

Rockstroh T, Abbott D, Hix K, Mook J (2013) Additive manufacturing at GE aviation. Industrial laser solutions (Iss. Nov/Dec), pp 4–6

Syed WH, Pinkerton JA, Li L (2006) Combining wire and coaxial powder feeding in laser direct metal deposition for rapid prototyping. Appl Surf Sci 252:4803–4808

Wholers T (2011) Growth of additive manufacturing and future opportunities and forecast. In: Additive manufacturing, Technology Roadmap for Australia, pp 10–14, p 38

Yoshimoto E, Kondo S (2012) Wing vein patterns of the Hemiptera insect Orosanga japonicus differ among individuals. Interface Focus 2:451–456

Zhang YT, Alber MS, Newman SA (2013) Mathematical modeling of vertebrate limb development. Math Biosci 243:1–17

TriVoc

Robotic Manufacturing for Affecting Sound Through Complex Curved Geometries

Dagmar Reinhardt, Densil Cabrera, Marjo Niemelä, Gabriele Ulacco and Alexander Jung

Abstract Complex curved surfaces posit challenges for manufacturing, but become more available through the range of toolpaths that come with 6-axis robotic fabrication. In this chapter, we present an in-progress report that explores the way in which an industrial 6-axis robot can become an interdisciplinary research tool that produces space that is both immediate and responsive. We link a robotic code directly to acoustic equations, so that in a reverse engineering process, kukalprc and robot reachability give boundary conditions for the consecutive design process. The chapter discusses a framework in which the robot is first used as subtractive manufacturing device for cutting an acoustically performative space, and indicates future research into the potential of a robotic assessment of complex geometries and the resulting acoustic performance. Through integration of acoustic behaviour and robotic fabrication parameters, the production of a space with three distinct 'sound colorations' becomes possible. Furthermore, future research is outlined whereby the robot acts as both hand and head: shaping an environment as both input and output device.

D. Reinhardt (✉) · D. Cabrera · M. Niemelä
The Faculty of Architecture, Design and Planning, The University of Sydney, Sydney, Australia
e-mail: dagmar.reinhardt@sydney.edu.au

D. Cabrera
e-mail: densil.cabrera@sydney.edu.au

M. Niemelä
e-mail: marjo.niemela@sydney.edu.au

G. Ulacco
AR-MA, Sydney, Australia
e-mail: gabriele.ulacco@ar-ma.net

A. Jung
Reinhardt_jung, Architecture and Design in Theory and Practice, Frankfurt, Germany
e-mail: jung@reinhardtjung.de

W. McGee and M. Ponce de Leon (eds.), *Robotic Fabrication in Architecture, Art and Design 2014*, DOI: 10.1007/978-3-319-04663-1_11,

163

Keywords Parametric design · Robotic fabrication · Curved surface geometries · Acoustic integration · Sound concentration

1 Introduction

The deployment of mathematical principles for the generation of complex curved geometries is a foundation of computational design methodologies, specifically in the area of generative and parametric design, by use of Rhino/plugin Grasshopper. The way in which these principles are adopted for design becomes critical at the moment when processes of construction and sound performance are aligned. While design and adaptation criteria in parametric variations and acoustic analysis allow multiple variations, the fabrication and manufacturing of complex curved surface geometries, including spheres and hyperboloids, remains a challenge. What are the advantages of directly interfacing acoustic performance and robotic fabrication? Can a reverse engineering of the design process—with direct linkages between acoustic equations and robotic toolpath—produce an advanced design process, and a deeper understanding of relationships between space, robot workspace, and sound? In which way can design, analysis, and fabrication software be interlinked?

This chapter discusses the research project 'TriVoc', termed so because a triad of three ellipsoids produces a harmonic inflection of human voices. Their sound reflection is manipulated through intersecting, curved geometries to interact with the sound within it, transforming nondescript sound such as speech/vocals in order to bring out a tonal character. The research extends previous work on parametric design of complex surface geometries and resulting acoustic performance capacities into the area of constructing acoustic performance through robotic manufacturing (Fig. 1). We are reporting on the work in progress, in which an acoustic equation is used as basic design principle for the ellipsoidal geometry of three intersecting spheres, and then mapped onto the reach envelope of an industrial robotic, so that the phenomenon of sound inflection can be produced through space.

The project explores this design through a framework for a reverse engineering process—from mathematical principles for acoustic performance (Matlab), made available as machine code (RhinoPython/GH), mapped onto robotic manufacturing (kukalprc), so that the design is immediately informed by fabrication constraints, and a capacity of actual (acoustic) performance (Comsol Multiphysics)—effectively establishing seamless shifts between the different disciplinary areas. Through this, the mathematical equation for an ellipsoid that reflects sound becomes the code for robotic fabrication, bypassing both generative design variations and acoustic simulations that have been part of a standard iterative design process.

In addition to an extensive use of 6-axis robotic fabrication procedures through direct programming—from acoustic equation to machine code—the robot is

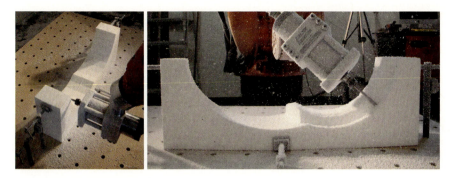

Fig. 1 TriVoc-triad space produced through robotic fabrication

further considered for its potential immediate response (sensing and continued adaptation of acoustic data). In doing so, we are testing architectural space as permanently impacted through material processes and sound.

2 Ellipsoidal Curvatures: Geometries that Affect Sound

In the context of audio-acoustics, domes, spheres, hyperboloids are commonly known to pose problems of sound concentration due to their curved geometries (Vercammen 2013). The sound concentration is a phenomenon that arises in circular spaces (circular ground plane/centre based sound source), when sound rays meet the wall at same time, are reflected, and converge in the form of a soundwave upon the original source position. The resulting sound concentration at centre point can adversely affect speech intelligibility and audience experience, and is perceived as disturbance, specifically in spaces used for temporal arts performances. Previous research has shown that these complex circular geometries can be adapted to an acceptable acoustic performance through iterative design loops that link parametric design (variations in Grasshopper), structural engineering (Strand7), and acoustic simulation (ODEON) (Reinhardt et al. 2012, 2013).

A systemic use of parametric design can also suggest surprising acoustic phenomena resulting from complex spatial geometries. For example, whereas circular geometries may complicate speech intelligibility, ellipsoidal spaces or vaults can create a particularly interesting case of sound concentration, because instead of a single centre, ellipsoids have two foci. This causes sound to be reflected in a very different manner, namely, all sound originating at one focal point is concentrated on the other focal point (Fig. 2). This effect is recognised as enhanced understanding of whispered sound that travels along the vault's surface. The geometry that is the base for this phenomenon has been widely documented (Kircher 1966; Langhans 1810; Cremer 1960). From a signal processing perspective, the resulting space can be thought of as an acoustic filter.

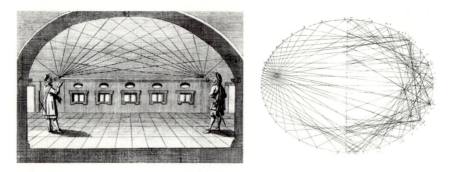

Fig. 2 Hearing space. Sound concentration in vault/ellipsoidal geometries

The acoustic behaviour of complex curved surfaces is a challenge that can be answered through robotic manufacturing as new mode of constructing space. Consequently, this research reviews the unexpected acoustic properties that result from focusing form. As part of wider investigation, the research will later include a matrix of geometric forms, which are investigated for their particular frequencies of sound, but at this particular stage, the research continued previous approaches to intersecting spheres and domes, and extended these to the present work: a series of intersecting ellipsoids. By interfacing parametric design process, structural analysis and acoustic analysis, and fine-tuning parameter (of dimension, distance, height, curvature, focus point, sound source and audience position), the acoustic behaviour of performance space in ellipsoidal structures can be generated, controlled, fabricated, and evaluated after completion through a head-model (Fig. 3). Through integration of robotic machine path, acoustical framework, and generative design, a compelling and distinctive geometry can be produced; a space that enhances acoustic performances/experiences by delivering a combination between beauty and sound reflection: an interior soundscape.

3 TriVoc: Concept of a Triad Space with 3 Pitches

Conceptually, the project envisions the robotic fabrication of a space based on a controlled sound reflection. The space is formed by three intersecting ellipsoids that produce a three-pitch chord, known in music as 'triad'. These ellipsoids have both a shared and individual geometrical aspects: they share foci, but are based on different radii, and different reflected path lengths between foci. Each pitch comes from the acoustic interference between the direct and reflected sound paths between pairs of foci, which form a 'comb filter' or harmonic series transfer function spectrum. By tuning the physical geometry of each ellipsoid, pitches are chosen within the range of maximum human pitch sensitivity. This produces a three-person conversation space that changes the tonal character of voices

Fig. 3 Measuring acoustic response of a reverberant space using a head and torso simulator

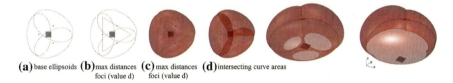

(**a**) base ellipsoids (**b**) max distances (**c**) max distances (**d**) intersecting curve areas
foci (value d)   foci (value d)

Fig. 4 A Triad Space generated through 3 intersecting, similar ellipsoids that reflect 'sound' equally (**a–d**), and manipulated to different pitches

('coloration' in architectural acoustics). This tri-focal triadic space brings the effect to the fore in a distinctive way. The base geometry was set up to follow a system that is adaptable through guiding parameters; by (1) definition of three distinct frequencies that are the result of particular individual curvature of the spheres, and by (2) definition of distance between foci of the ellipsoids that reflect sound back. Departing from equal frequencies of 500 Hz (5:5:5 ratio), the project focused on a triadic space tuned to 400, 500 and 600 Hz (4:5:6 ratio). With co-foci arranged in an equilateral triangle, the major and minor radii of each ellipsoid are derived (Fig. 4).

The fine-tuning through different frequencies results in a person situated at one of the foci to be able to hear the voice of a second person filtered by one of the pitches, and the voice of the third person filtered by another pitch. Thus, for one listener the space is characterised by a major third, for another by a minor third, and for third listener by a perfect fifth. Each person perceives their two partners' voices in a different manner. A person situated at one of the foci can hear the voice of the second person filtered by one of the pitches, and the voice of the third person filtered by another pitch, but the third pitch is audible between the second and third people only. For this effect to come alive, the translation between sound equation, control over geometry, and precise manufacturing is crucial. For that reason, we developed a framework that could integrate the acoustic base equation through to machine code.

4 Acoustic Base Equations for Ellipsoids in TriVoc

Acoustic parameters that inform or consolidate complex geometries can be embedded through multiple integration of design, analysis and optimization processes. Commonly, this requires a number of operations between different computational packages: simulation software (Comsol Multiphysics), plugin (McNeel Rhino/Pachyderm), and mathematical equations (Matlab). In contemporary design-led research, such iterative sequences have been deployed to improve acoustic performance by sound in relation to source and target positions (Williams et al. 2013; Reinhardt et al. 2012). Specifically the generic shape (ellipsoidal, spherical or dome structures), or the micro-patterning of surfaces can be altered; by adapting the height, dimension and centre point; or by adapting through faceting, which consequently improves the acoustic behaviour of the performance space.

Yet while such optimisation and adaptation process requires complex transfers between the different computational software, this normally does not include the specific fabrication constraints related to complex curvatures, which require specific degrees of freedom in the tooling path. This gap set the challenge for the work on TriVoc, where we reverse engineered the typical process. Instead of the usual design path of applying an acoustic simulation to the 3D model (Rhino/Grasshopper), receiving acoustic feedback and producing further design iterations, we synchronized the acoustic equation with the robotic workcell and in doing so allowed project constraints to become determining drivers for manufacturing desired surface curvature. The basic geometric modelling in this project was developed in Matlab (a computational and signal processing environment widely used in acoustics) which generates x, y, and z coordinates for three ellipsoidal intersecting bodies (Fig. 5). While we tested different variations of scripting ellipsoids/circular geometries, it became quickly apparent that a conventional generative design path—with design iterations limited to Rhino/Grasshopper—would not provide sufficient information for fabrication (Fig. 6).

At this stage, the equation allowing us access to identify the first relationship between geometrical variation and resulting sound became a determinant for the precise curvature of the ellipsoids. The fundamental frequency, f (in Hz), of the comb filter produced by the interference between direct and reflected sound between ellipsoid foci is calculated by following equation:

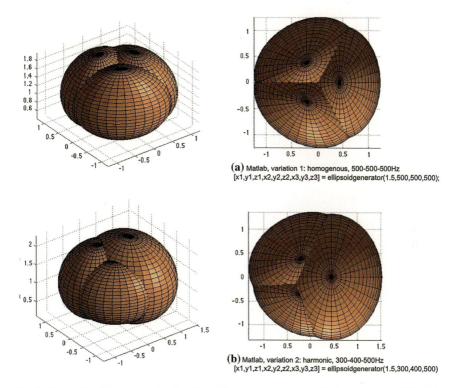

(**a**) Matlab, variation 1: homogenous, 500-500-500Hz
[x1,y1,z1,x2,y2,z2,x3,y3,z3] = ellipsoidgenerator(1.5,500,500,500);

(**b**) Matlab, variation 2: harmonic, 300-400-500Hz
[x1,y1,z1,x2,y2,z2,x3,y3,z3] = ellipsoidgenerator(1.5,300,400,500)

Fig. 5 Generating Matlab function for ellipsoids that form variation (*1*) homogeneous 500–500–500 Hz space, and (*2*) harmonic 400–500–600 Hz space, generates curve equation

$$f = \frac{c}{2a - 2\sqrt{a^2 - b^2}}$$

here, c is the speed of sound (in m/s), a is the ellipsoid's major radius, and b is the minor radius (both in m). Note that although an ellipsoid has two minor radii, they must be identical for the ellipsoid surface to produce a constant time delay for sound reflected off its surface traveling from focus to focus. Based on the input parameters of distance between foci, and the respective tuning frequencies of each ellipsoid, this function can be deployed towards different scenarios: three identical intersecting ellipsoids, or the more complex scenarios pursued here, a space that delivers a musical triad. By varying the distance between foci, and the respective tuning of three intersecting ellipsoids (in Hz), a number of geometrical solutions can be derived that look and sound very different to each other.

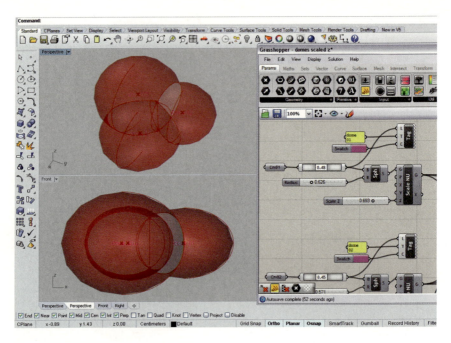

Fig. 6 Parallel parametric modelling

5 Reverse Engineering: Matlab to RhinoPython

This initial definition of acoustic thresholds for ellipsoids needed to be made
available for both generative design and fabrication methodologies. Here, we used
a mathematical equation for ellipsoids in order to produce an efficient intersection
between geometries of fabrication and acoustic performance, in turn controlling
the sound behaviour through mathematical language: a framework for computation
of sound through space. In order to enable integration, the acoustic geometries
defined by Matlab were taken into RhinoPython as base equation for curvatures
and reverse engineered for the integration into robotic toolpath in kukalprc. The
resulting project system is based on the following GHPython code that generates
the individual ellipsoids:

```
import rhinoscriptsyntax as rs
import math

#function [x1,y1,z1,x2,y2,z2,x3,y3,z3] = el-
lipsoidgenerator(d,f1,f2,f3)
# generates the coordinates of three ellip-
soids with co-located foci.
#
# d is the distance between foci in metres
#
# f1, f2 and f3 are the fundamental frequen-
cies of comb filters formed by
# the interaction between the direct and first-
order reflected sound
# between ellipsoid foci. Each ellipsoid can
be tuned differently (if desired)

c = 344 # speed of sound in m/s


elevation = 1200 # elevation of the equator
in mm
n = 32 # used in ellipsoid function to control
the number of coordinates generated

c = c*1000
```

```
delay1 = 1/f1
delay2 = 1/f2
delay3 = 1/f3

extradistance1 = c * delay1
extradistance2 = c * delay2
extradistance3 = c * delay3

# Displacement of ellipsoid centres from
origin
R = math.pow(3,0.5) / 6 * d

# major radii
a1 = (extradistance1 + d)/2
a2 = (extradistance2 + d)/2
a3 = (extradistance3 + d)/2

# minor radii

b1   =   math.pow((math.pow(a1,2))   -
(math.pow(d/2,2)),0.5)
b2   =   math.pow((math.pow(a2,2))   -
(math.pow(d/2,2)),0.5)
b3   =   math.pow((math.pow(a3,2))   -
(math.pow(d/2,2)),0.5)
```

This enables a control over the various dimensions and distances in the ellip-
soids, effectively controlling the vault lines that result in sound reflection. In that
manner, the pitch in each of the sphere parts, as much as their relation towards
each other can be manipulated. More importantly, this intersection between
Matlab, RhinoPython/GH enables control over the sound pitch and spatial design
(Fig. 7), but also supports a workflow between design and robotic fabrication.
Constraints of workcell and project dimensions can be directly tested in kukalprc
(Fig. 8) in order to fit or adapt the project to the space a robot can requires.

6 Defining Constraints: Embedding Geometries in Robotic Reach Envelope

In terms of fabrication, the project extends previous research into complex curved
geometries by exploring two aspects through robotic manufacturing: the aspect of
subtractive space shaping (by carving into solids) and the toolpath accessibility
constraints (by developing design through machine code). Through multiple level
integration between Matlab, Grasshopper, and kukalprc, work context are
combined.

TriVoc is designed as 'subtraction' of architectural space from a simple solid,
understanding the intersecting spaces as virtual domain, the area of sound to be
carved out. The robotic manufacturing process is thus designed to subtract layers
of material from a composite block of stacked standard Styrofoam sheets (EPS
expanded Styrofoam M Grade, 152 mm thickness) until the target surface is

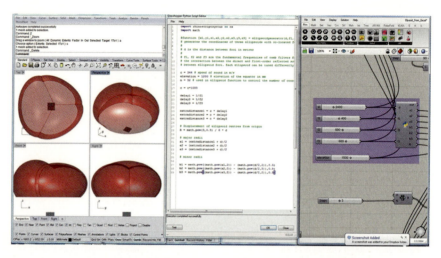

Fig. 7 Triad in RhinoPython|GH code, based on Matlab equation, 400:500:600

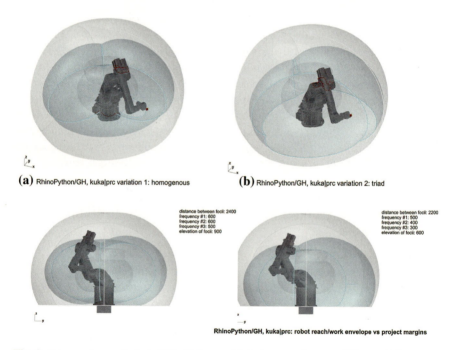

(**a**) RhinoPython/GH, kuka|prc variation 1: homogenous

(**b**) RhinoPython/GH, kuka|prc variation 2: triad

distance between foci: 2400
frequency #1: 600
frequency #2: 600
frequency #3: 500
elevation of foci: 900

distance between foci: 2200
frequency #1: 500
frequency #2: 400
frequency #3: 300
elevation of foci: 600

RhinoPython/GH, kuka|prc: robot reach/work envelope vs project margins

Fig. 8 Comparative analysis in RhinoPython and kuka|prc, equal versus different pitch

reached (in this case the ellipsoidal sound space), effectively leveraging embedded ellipsoids and their inherent acoustic performance. Toolpath freedom provided by the robot is essential for the project. While fabrication processes in architecture often rely on parts that are fabricated separately (Kilian et al. 2007), this space is

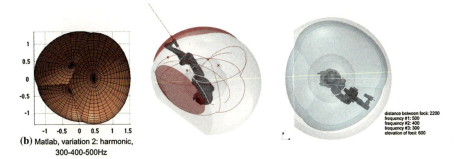

(b) Matlab, variation 2: harmonic,
300-400-500Hz

Fig. 9 Multiple level integration: Matlab, Grasshopper, kukalprc and work context are interfaced

carved out as a totality, meaning that the degrees of freedom (DOF) are fully exploited along the robot's 6-axes. Since the project required multiple directions for a complete reach envelope in order to carve the project from the material solid, the alignment between available form and available toolpath was critical (Fig. 9).

A critical step in robotic manufacturing is the identification of the workcell. This requires a differentiation of 'regions' of geometrical envelope and robotic reach; a separation between desirable geometry and possible fabrication (Menges 2013). Further constraints included the angle of milling tool, overhead rotation, and the kinematics of a 6-axis reach of a standard KUKA KR60-3 equipped with a milling tool as end-effector. To achieve a successful integration, the project required us to identify the functional constraint boundaries of the robot, i.e. the robotic reach envelope, and to set this in direct relationship with the 'producible' constraint boundary—the required acoustic/spatial geometry of the triad space. Based on the project geometry derived from Matlab/RhinoPython|GH code, the project went through a series of investigations that mapped constraints and intersected between acoustic shape, robot path, work envelope, and work site (Fig. 10). The specific robotic path for the 6-axis KUKA KR60-3 was generated by kukalprc interfacing the acoustically defined geometries on a global level. This mapping of constraints ensured the adequate integration of the mechanical process, especially where problems became apparent and could be corrected (e.g. project volume penetrating available work envelope). For the carving of ellipsoids, the rotation of tool axis, moving the tooltip along the EPS Styrofoam surface, and making sure that drill positions could be delivered without the robot passing through its own machinery (base, power cables) were taken into account. The dimensions of robot reach envelope were further specified by endpoint of industrial roll plane in the KUKA60-3 (extended by end effector, in this case a 4 KW milling spindle, Fig. 11).

The 6-axis robotic milling was especially useful for the project requirements since (a) the undercut required by the acoustic ellipsoids (datum beyond equator) cannot be produced in a single process through other forms of manufacturing with fewer degrees of freedom and (b) a material finish can be achieved that allows for

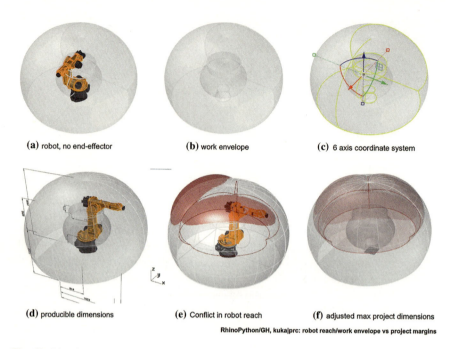

(**a**) robot, no end-effector   (**b**) work envelope   (**c**) 6 axis coordinate system

(**d**) producible dimensions   (**e**) Conflict in robot reach   (**f**) adjusted max project dimensions

RhinoPython/GH, kuka|prc: robot reach/work envelope vs project margins

Fig. 10 Mapping constraints: ellipsoids, robot reach envelope, default/conflicts, and adapted project

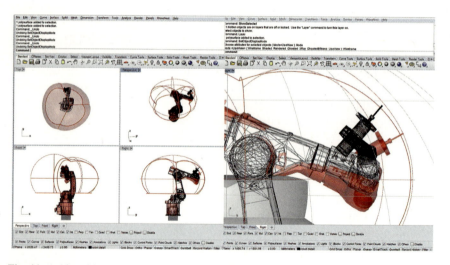

Fig. 11 Achieved level integration: access centre point through a variety of robot positions

immediate acoustic testing without further work on the surfaces. The definition for radii, focus points, and curvature for the three intersecting ellipsoids was then

Fig. 12 Robotic workspace, intersecting ellipsoidal curves

adjusted on a local level to fit the robotic work envelope and design variations developed through Rhino/GH. Then, the subtractive toolpath was finalised, so that material layers could be removed to shape acoustic space(s). This embeds also the potential for a continued modification of the environment (Oxman 2013), whereby the robot carves, listens, and continues to carve—a process whereby robotic capacities can be leveraged as both an input and output device.

7 Scaled Prototype and Acoustic Simulation

As a work in progress, TriVoc transfers mathematical equation to machinable geometry and acoustic behaviour in 1:1 prototype. In preparation, we ran several test versions, including material acoustic tests (which confirmed the 'acoustic fitness' of the selected Styrofoam (density) (normal incidence impedance tube testing of absorption coefficient showed that it reflected at least 90 % of incident energy across the relevant frequency range), and segment prototype tests to confirm the approach to robotic toolpath (Fig. 12). In preparation of a full scale model, and prior to fabricating a 1:1 prototype, we continued in a first step at producing a functioning (acoustically proficient) prototype at 1:4, which required scaling in shape, and impacted on sound, but fundamentally maintained the most critical parameters of control over robotic toolpath: control over curvature and availability of resulting pitch for physical analysis.

In acoustic terms, the scale model for acoustic testing is simply done by shifting the measurement by two octaves (e.g. 4 kHz in the scale model is used to simulate the acoustic response at 1 kHz full scale). The shift in scale also requires a change of sensing technology, so impulse response measurements will be done between the three foci, using a small loudspeaker and microphone. The impulse response represents the full pattern of acoustic reflections. By decreasing its speed by a factor of 4 and convolving it with anechoic speech, it is possible to listen to an approximation of the full-scale model. This convolution can be done in real time

Fig. 13 Robotic workspace on site, material prototyping

(with a latency of only a few milliseconds), so that the sound of the three-way conversation in the envisaged full-scale space can be experienced. The spectral effect of the reflections can be examined by Fourier transform of the impulse responses. As a sub-stage to avoid material waste and produce first results with the potential of optimisation after acoustic feedback, the TriVoc project was then reconfigured with decreased radius size of ellipsoids, and individually manufactured elements (Fig. 13).

8 Conclusion and Future Research

The project discussed a framework for establishing pathways for robotics (reachability, fabrication, application), supported through interdisciplinary approaches (acoustics, parametric design) to its mathematical and geometrical conditions. As a result, a multiple level integration enables control over complex curved geometries, design variations, robotic reach and toolpath, and site constraints. By embedding a mathematical language, the research discussed the way in which the equation for an acoustic performance, integrated into the complex curved geometry of ellipsoidal spaces, could be directly linked to robotic fabrication code. In a context of research-based design, this is further important because it posits a strong argument for the 'construction' of architectural design, by deploying mathematical principles that are shared in interdisciplinary discourse.

As a progress report, a number of following stages shall be indicated as continuation of this research. On a general level, future research can be undertaken through public communal spaces that are reorganised when former use abandoned (churches, industrial sites, libraries). By producing large concave/convex modules with embedded faceting (in relation to double curved surfaces of gothic/baroque architecture), certain frequencies of sound can be accentuated. For those, acoustic improvement can be provided by the insertion of custom acoustic panels.

More importantly, and directly linked to this framework, is a research planned after completion of 1:4/1:1 robotic manufacturing. In a second stage, the research will run through an acoustic simulation. In the full-scale fabrication, the acoustic response can be explored more precisely in relation to the human form. The analysis will proceed as human audience response, and though robotic sensing. A dummy head replaces then the drilling tool on the robot, allowing the effect of the head's presence and orientation to be examined through a large number of auto-mated measurements. Furthermore, the sound source in the acoustic testing will be a head and torso simulator, which incorporates a mouth. This more sophisticated approach to testing allows us to optimize the design for the directivity of the human voice and the directional sensitivity of hearing.

In a third stage (projected), the robot continues to carve, in a continued mod-ification of its environment (Oxman 2012). This requires another segmentation of the ellipsoidal spheres, whereby sound will be manipulated through a patterning of the interior surfaces, again varying the sound for each of the person listening to another inside the volumes. In this manner, robotic fabrication can aid in the design and fabrication of complex spatial arrangements with distinctive acoustic performance that is both immediate and responsive.

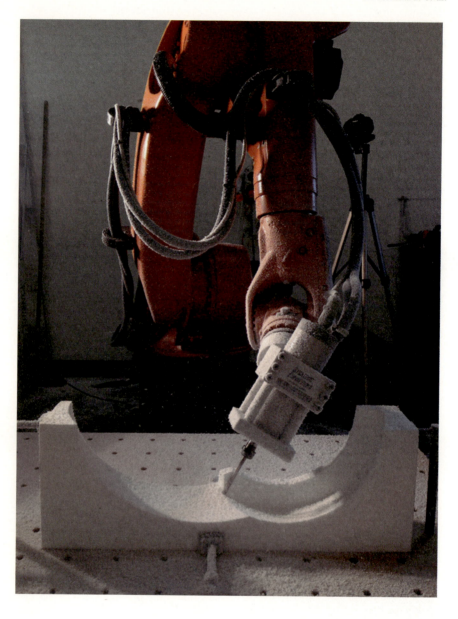

References

Cremer L, Mueller L (1982) Principles and applications of room acoustics. Applied Science, New York

Keating S, Oxman N (2012) Robotic immaterial fabrication. In: Brell-Cokcan S, Braumann J (eds) RobArch: robotic fabrication in architecture, art and design. Springer Wien, New York

Kircher A (1966) Phonurgia Nova, Broude, New York

Langhans CF (1810) Ueber Theater, Julius Eduard Hitzig, Berlin

Menges A (2013) Morphospaces of robotic fabrication: from theoretical morphology to design computation and digital fabrication in architecture. In: Brell-Cokcan S, Braumann J (eds) (2012) RobArch: robotic fabrication in architecture, art and design. Springer Wien, New York

Pottmann P, Asperl A, Hofer M, Kilian A (2007) Architectural geometry. Bentley, USA

Reinhardt D, Martens W, Miranda L (2012) Acoustic consequences of performative structures—modelling dependencies between spatial formation and acoustic behaviour. In: Achten H, Pavlicek J, Hulin J, Matejdan D (eds) Digital Physicality—Proceedings of the 30th eCAADe Conference, vol 1. Czech Technical University in Prague, Faculty of Architecture, Czech Republic, pp 577–586, 12–14 Sep 2012

Reinhardt D, Martens W, Miranda L (2013) Sonic domes: interfacing generative design, structural engineering and acoustic behaviour. In: Stouffs R, Janssen P, Roudavski S, Tunçer B (eds) Open systems (CAADRIA 2013), pp 529–538

Vercammen M (2013) Sound concentration caused by curved surfaces. J Acoust Soc Am http://scholar.qsensei.com/content/1vs7mb

Williams N, Davis D, Peters B, De Leon AP, Burry J, Burry M (2013) FABPOD: an open design-to-fabrication system. In: Stouffs R, Janssen P, Roudavski S, Tunçer B (eds) Open systems (CAADRIA 2013), pp 251–260

Performative Tectonics

Robotic Fabrication Methodology Towards Complexity

Philip F. Yuan, Hao Meng and Pradeep Devadass

Abstract This chapter addresses the challenges facing current contemporary practices in the development and execution of complex designs by discussing the integration of computational design and digital fabrication methods in the project Light-Vault. The project shows the development of a vault created through the aggregation of several dissimilar component sin which the interior volume is carved out, leading to the formation of ruled surfaces. Porosity of the component is parametrically designed through a genetic algorithm controlled by multiple fitness criteria. In parallel, the project explores and implements the potential of robotic technology by developing personalized robotic tools and production techniques to quickly shape volumes with a hot-wire cutting process. Algorithms are developed to ensure design thinking and fabrication procedures are simultaneously developed in a, non-linear, parallel performance based process. This cumulative cohesive process between advanced digital and physical computation methods is translated through a full-scale built pavilion.

Keywords Ruled surfaces · Genetic algorithm · Hot-wire cutting · Robotic fabrication technology

P. F. Yuan (✉)
College of Architecture and Urban Planning, Tongji University, Shanghai, China
e-mail: philipyuan007@gmail.com

H. Meng · P. Devadass
Archi-Union Architects, Shanghai, China
e-mail: hydemeng@gmail.com

P. Devadass
e-mail: pradeep.devadass@hotmail.com

W. McGee and M. Ponce de Leon (eds.), *Robotic Fabrication in Architecture, Art and Design 2014*, DOI: 10.1007/978-3-319-04663-1_12,

1 Introduction

1.1 Towards Morphological Complexity

Today's contemporary practices are currently experiencing interests in complex design language, not only in terms of visual sensuality but also in the technicality of the system. Yet, they are constrained by limitations, mainly cost and time. The morphological development by computational methodologies in the current era leading to non-Euclidean geometrical forms have been vastly exhibited in modern architecture showing highly intricate complex visual form (http://www.jstor.org/discover/10.2307/985254?uid=3737800&uid=2&uid=4&sid=21103014331677).

The current rise in implementation of technology and fabrication tools in design encourages designers to redefine building elegance and efficient structures through form finding methods (Block 2009). Research on form development does not restrict itself to architecture but also extends to other sectors of design industry (largely in interior, fashion and product design). According to Peter Zellner, "architecture is recasting itself, becoming in part of an experimental investigation of topological geometries, partly a computational orchestration of robotic material production and partly a generative, kinematic sculpting of space" (Burry 1999). This form of design methodology using computational tools and algorithmic approach leads to a generative design process resulting in the complication of the system displaying graphical intricacy. The complexity also extends to experimental geometry, fabrication techniques, material intelligence and various other dynamics of design development.

The project Light-Vault questions rectilinear form by demonstrating an organic morphological design process using freeform development defined by algorithms controlled by parameters. In nature, growth is governed by a set of instructions influenced by the external stimuli. Cells are generated and regenerated, leading to an individualistic and characteristic morphological development, making it challenging to realize the intelligence behind the logical approach easily perceived by the external morphology (Vogel 2003). This systematic methodology in nature is translated into the project using computational methods integrating performance of the system, materiality, fabrication constraints, component logistics and assembly process.

1.2 Fabricating Complexity

Fabrication of complex geometries/double curved surfaces through manual methods has always been challenging mainly due to the lack of skilled labor and human physical limitations; however, with advent of digital fabrication the erection of these structures is easily conceivable. Precise manufacturing, reduction of human labor, optimization of resources, cost effectiveness, and quick assembly processes are some of the major advantages distinguishing digital manufacturing technology as the most sought out in current modern practice. CNC milling is popular in subtractive methods, but an extensive process. Therefore, the requirement of a less time consuming procedure is found to be crucial.

Hot-wire cutting process has been commonly used in the foam industry for making sculptures, extraction of raw materials, etc., by carving out large volumes in short duration of time. The process can be related to stereotomy (art of stone cutting), simulating rapid and efficient cutting of natural stone with a diamond-wire saw (Asche 2000). Polyurethane foam's inherent properties of a low melting point and softness facilitates the production of desired shapes effortlessly, making it one of the most popularly available materials used widely in creation of moulds, sculpture and various other experimental formworks. The processing time of hot-wire cutting methods are comparatively faster than other manufacturing methods, but can vary depending on the detail and size requirement of the end product. For obvious reasons, to improve existing subtractive method in order to achieve complex designs, hot-wire cutting is digitally executed with the help of 6-axis robots.

In Light-Vault, the project studies and adapts the fabrication methods from past examples like RDM Vault by Hyperbody and Rippmann Oesterle Knauss/ETH Zurich, and, ReVAULT by Maciej Kaczynski, Wes McGee, and Dave Pigram at University of Michigan Taubman College, and pushes the limits of the above processes in a limited time frame, exploring optimization techniques through parametric designing of components where fabrication procedure and its constraints are considered as one of the main parameters. Optimization techniques are occurring during every stage of the process, from the initiation of design development, to the manufacturing procedure through a controlled digital environment. From design to fabrication, the process is completely carried out at Archi-Union architectural office, Shanghai, one of the few commercial practices around the world to house a 6-axis, Kuka Robot (KR 30-3). The incorporation of advanced digital fabrication methods and use of robotic technology in commercial design affords designers complete control to experiment on full scale built prototypes.

2 Vaulting Systems

Vaulting systems have been widely used, mainly in the roof systems of gothic churches. Recent past examples include the asymmetric barrel vault of Fronton Recoletos, one of the most innovative structures created by Eduardo Torroja (Acland 1972), and Eladio Dieste's Gaussian vault which utilizes the ground-breaking construction method in which a double-curve catenary arch (which resists buckling) is formed out of a single-shell structure (Dieste 2004). A Gaussian vault is designed and reconstructed in the project Light-Vault for the pavilion at 10th anniversary exhibition of Archi-Union, to test various aspects of structural engineering, focusing on the limits of span, thickness and porosity of the structure.

The vault is located in a closed exhibition space surrounded by 3 walls to display the exhibition panels. The vault is designed to cover a span of 6.6 m long, 4.45 m wide and 3.52 m high. To avoid obstructing the movement of the visitors, one of four supporting bases of the vault is placed on top of the inner display wall. One of the main considerations of the design was to develop a methodology based on mathematical rules to test the combination of two distinct ideas of Torroja and Dieste by creating a thin, asymmetrical, catenary arch vault.

3 Methodology Overview

In design practices, the decisions on execution of a project, in a linear process, are developed at the final stage of design phase. A parallel thinking methodology is developed to combine design development and fabrication process. Execution of a design has always been an isolated unparalleled process developed at the end of the design stage. Evans (1997) explains the inevitable gap between drawing, the medium of design, and the final outcome in architecture. This distinct unconnected procedure leads to a problematic translation from digital environment into reality. Development of a methodology from design to fabrication through the integration of technology in a controlled digital environment is established in the project Light-Vault. A performance based design approach is the key focus in developing the methodology. Also, evaluation is normally generated during the end of the design stage leading to the absence of continuous feedback for the improvement of the design output. Genetic algorithms are developed for optimization of a system, imitating the process of natural selection (Mitchell 1996) (Fig. 1).

The algorithm develops a cohesive process, running a parallel design and fabrication procedure through sequential evaluations to ensure an optimized design output. The design evolves through a generative process by analyzing and evaluating at every stage of the procedure, which results in significant improvement in the output. This interconnection between various aspects of design development, allows the designer to manually decide on specific factors which not only results in a better design but also a cost effective production process.

4 Vault Development

In principle, a vault comprises structural stability but due to the placement of bases on two different levels, the developed algorithm critically calculates structural stability using Thrust Network Analysis (TNA) for solving the horizontal and vertical forces in equilibrium of the vault through the structural analysis tool, Rhino-VAULT, developed by BLOCK Research Group at ETH, Zurich. The tool uses evolutionary structural optimization to provide a more interactive design approach

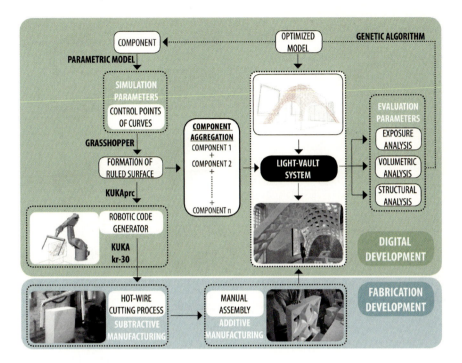

Fig. 1 Methodology overview

(http://alphard.cpm.aca.mmu.ac.uk/combib/combib.html). Structural requirements include consideration of the local thickness of the vault formed through extrusion of each component and orientation of the tessellation in which the contact faces of neighboring components want to be aligned to the local force flow (Rippmann and Block 2011). In order to solve this problem a design is generated out of non-planar components in the form of NURBS surfaces. The surfaces are planarized with minimal deflection, taking into account the considerable thickness in correspondence with reality. The entire system is optimized using parametric physics engine tool (the Grasshopper plugin Kangaroo, developed by Daniel Picker) to ensure the maximum tolerance between connecting points does not increase over 2 mm in reality (Fig. 2).

Due to the large span, the design form is optimized into discrete developable quadrilateral surfaces of optimum sizes through tessellation which depends on the curvature of the surface at that region using the developed algorithm. The surfaces are later developed into components by adding thickness of the raw material blocks. In order to achieve an efficient structure, each component of the vault system is rearranged with respect to its volume (which in turn results in weight of the unit); units are designed to be heavier at the base and lighter as it moves upwards in the vaulting system. The formation of components through tessellation in the final vault design form result in a top-down approach.

Fig. 2 Structural formation of the vault (*left*); optimization of vault components (*right box*)

5 Component Development

5.1 Evaluation Parameters

The acknowledgement of performance in a building according to an ever changing environment establishes flexible systems benefitting the user experience (Kolarevic and Malkawi 2004). The integrity of design is developed through a series of tests dependent on external dynamics, thereby influencing performance. In order to test this performance, improving the quality of a design, fitness criteria are required to be considered. The continuous uninterrupted flow in visitors' movement and clarity to display exhibits are considered important functions of an exhibition space. Placing a massive structure, such as the vault, in an exhibition space with the least interference and disturbance becomes a challenging task. Factors such as light (required for emphasizing exhibits) and transparency of the structure (to avoid obstruction the view of the visitor) become important aspects in development of the design. To achieve these factors, two evaluation parameters are considered:

- Exposure of light through the component.
- Visibility through the component.

Porosity is developed in the components to attain the above evaluation parameters by carving out an enormous volume through a defined process, giving rise to formation of ruled surfaces inside the component. This subtraction of a volume results in the edges towards the exterior appearing slender while still keeping a considerable volume intact, creating an illusion of the structure being of a lighter mass to the observer.

5.2 Formation of Ruled Surfaces

A ruled surface is generally formed by a continuous moving line in space which is guided by a curve. Here, two continuous polynomial closed curves are considered which project two straight lines. The intersection of these straight lines leads to the creation of points. These points results in the formation of the final doubly curved surface inside the component. These curves are further controlled by manipulating the control points of the each curve on the horizontal plane which are restricted by a boundary of the component. The boundary is formed through a minimum offset of 30 mm from the exterior surface to avoid any fragile thickness in the component that might lead to instability of the entire structure. The ruled surfaces are capped on either side, thereby forming the volume to be extracted from the component. Hence control points move only along the horizontal plane on the exterior opposite faces. The percentage of porosity which is extracted from each component is limited from 22 to 47 % of the total volume to sustain stability in the structure (Fig. 3).

5.3 Simulation Parameters: Control Points

The movement of the control points results in adjustments to the size and shape of the opening/porosity in the component. The amount of light through the component is measured using an exposure tool Geco, a Grasshopper plug-in developed by uto-Tools Group. The visibility through the component is measured through a Python script developed by the authors. Optimized positioning of the control points is controlled by a developed genetic algorithm utilizing the evolutionary tool Galapagos, a built-in Grasshopper plug-in. As the movement of points are on a horizontal plane, two genes which manipulate the X and Y coordinate values are considered to generate the position of the points. These values are limited in movement through a maximum distance to avoid any inter-collision of points or formation of irregular ruled surfaces. The complete control over the entire process is achieved through the parametric-genetic algorithm resulting in a performance based development in the formation of ruled surfaces (Fig. 4).

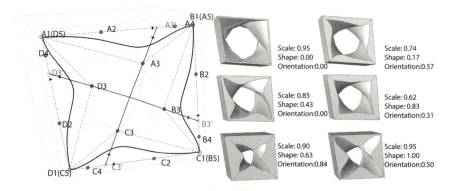

Fig. 3 Formation of a ruled surface using control points (*left*); formation of the void through different iterations by actuation of genes (*right*)

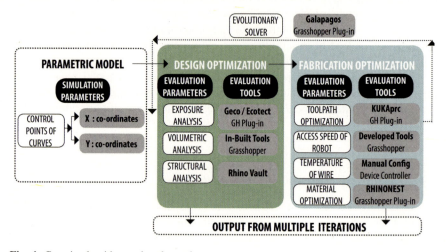

Fig. 4 Genetic algorithm and tools used

6 Translation of Digital Design into Physical Prototype

6.1 Software Development

The entire design component is developed in Grasshopper developed by David Rutten for the Rhinoceros 3D platform developed by Robert McNeel and Associates. A similar software approach based on the project RDM vault as discussed in "Processes for an Architecture of Volume" by Wes McGee, Jelle Feringa and Asbjørn Søndergaardis applied to generate customized toolpaths for robotic hot-wire cutting (RHWC) (Brell-Cokcan and Braumann 2012). Component sizes are

limited according to the readily available size of polyurethane blocks
(600 × 600 × 200 mm). Fabrication of components from readily available raw
material blocks is optimized by positioning and re-orienting the parts through
RhinoNEST, a Rhino Plug-in developed by TDM solutions in order to reduce
material wastage.

Ruled surfaces are generated through two polynomial continuous closed curves
in the digital model, however, to control the path of the robotic arm in cutting the
inner doubly curved ruled surface, two parameters are required to be defined:
position (mid-point) and angle of the wire with respect to that point on the surface.
Coordinates and projected lines at every point which form the ruled surface are
extracted. For the outer surface of the component, which in this case is a singly
ruled surface, coordinates at the mid-point and angle between curves is taken. To
increase efficiency in production, two different methods are developed respective
of the type of surface. The translation of this information from digital design
model into physical prototype through a sequential set of codes or instructions for
the robot are generated by robotic simulation software, Robots in Architecture's
KUKAprc. The software generates toolpaths to virtually test the motion of the
robotic arm and avoid any accidents in reality and allows for the user to manually
modify the robotic movement. The software is compatible with a parametric
plugin tool (Grasshopper) which extends the parametric interface in the modeling
software (Rhinoceros), allowing the designer to simultaneously integrate design
with manufacturing process.

6.2 Modification of the Robotic Arm

The heating mechanism of the metallic wire is suspended at one end of the wooden
frame and attached to the robotic arm on the other end which is controlled by an
electronically controlled circuit. Over time the metallic wire tends to lose its
tension by expansion due to overheating, resulting from the continuous electric
supply and friction of cutting. This electronic circuit contains a sensor to alert the
user during such discrepancies. Intense care is taken to make sure the ruled surfaces could be made through a single cut without any discontinuity between the
cutting process, leading to a smoother and faster method. The thickness of the wire
is taken into consideration and sufficient tolerance is given during the manufacturing process to achieve maximum accuracy (Fig. 5).

6.3 Production Techniques

Light-Vault develops fabrication methods based on previous research works on
robotic wire cutting at University of Michigan Taubman College. But unlike other
projects where the research extends to significant testing of different materials

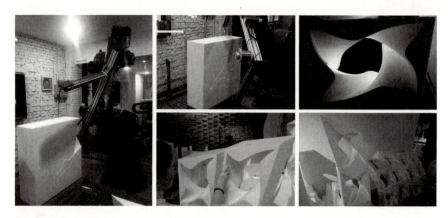

Fig. 5 Robotic fabrication and assembly process

using various cutting wires (Clifford 2012), Light-Vault limits and improves the
accuracy and quality of RHWC method by developing production techniques
based on speed of the robotic arm and temperature of the wire.

During production, there are several practical difficulties. The wire does not
evenly heat up throughout its entire length. The ends heat up more than the center
which results in the center of the wire cutting slower than the ends. If speed of the
cutting process in increased, the ends cut faster than the center of wire thereby
giving an arc shape to the wire and resulting in deformation of the component
towards the edges. Another problem encountered in the process is, if the hot wire is
simulated faster, the cut becomes coarse; if simulated slower, the wire starts to
melt the foam and the cut leads to an uneven surface. Thus an optimum speed of
cutting process and temperature of wire is found through various trial and error
methods. These parameters are largely influenced by the curvature of the component and density of the material. Hence, different densities are also selected and
tested. To increase accuracy and quality of the cut at corners and acute curvatures
of the ruled surface, profile curves are added to the digital model at that section.
An increase in the number of profile curves on the ruled surface is directly proportional to amount of accuracy and time taken for execution.

All inner ruled surfaces are closed and are formed at the center of the component this results in the robot cutting through the sides of the component to reach
to the center. To overcome this problem the metallic wire is inserted through the
center of the component by creating minute opening for the cuts starting position.

After the cutting process, the remaining central volume to be extracted from the
component is difficult to remove due to the formation of interlocking surfaces.
Hence the central volume is further cut into smaller pieces which can be removed
easily. Robotic manufacturing produces high precision components that are easier
to aggregate. Due to modularity of the components, the system becomes a simple
to assemble resulting in a smaller team of 5 in-house architects to execute the
entire structure in less than a day (Fig. 6).

Fig. 6 Light-Vault

7 Conclusions

This approach creates a new paradigm in methodology by achieving performance and fabrication through optimization.

7.1 Optimization of the Design Performance

Development of design and fabrication with a controlled parametric approach, continuous feedback by synchronization, and interlinking of various parameters through a single algorithm gives complete control to the designer and realizes flexibility in optimizing performance at every stage, leading to a significant improvement in the quality of the output.

7.2 Optimization of Fabrication Process

Optimization of design by incorporation of available raw material in fabrication of components results in a cost efficient method. A methodology where parameters such as accuracy of the size and shape, production quality, and time duration in the process are individually optimized (making the right choices and creating balance) can be exploited by commercial offices according to their requirements.

7.3 Limitations of Design Algorithm

Approximate evaluation results from the digital algorithm in relation to reality are a consequence of unrealized parameters not considered in the system, such as example fluctuating temperature of the wire due to friction. Optimization of design with these results is still questionable. The evolutionary solver (Galapagos) in the algorithm does not entirely facilitate multiple fitness criteria thereby leading to a limited development in the performance of the design.

7.4 Limitations of Fabrication Methods

There are numerous restrictions in the scale of production where the component needs to be of an optimum size due to the presence of a frame. Flexibility in manual methods is found lacking in digital tools (6-axis robots) due to defined and systematic interlinked approach. Producing closed geometries by inserting a wire through the center of the component and its alignment becomes a tedious process. The production of non-oriented geometries (Mobius strip) is unfeasible due to limited access of the 6-axis robots. Although hot wire process is a quick method in carving out large volumes, the fabrication of ruled surfaces with intricate details is still questionable.

Due to quick manufacturing capabilities, this process can be applied for contouring vast landscape forms through the use of self-transportable robots. There are continuous efforts to create frameless hot-wire cutting processes for producing more complex geometries.

References

Acland JH (1972) Medieval structure: the gothic vault. University of Toronto Press, Toronto

Asche J (2000) Tiefschleifen von granit. PhD thesis, University of Hannover, Germany

Block P (2009) Thrust network analysis: exploring three-dimensional equilibrium. PhD thesis, Massachusetts Institute of Technology, Cambridge, MA, USA

Brell-Cokcan S, Braumann J (eds) (2012) Rob|Arch robotic fabrication in architecture, art, and design. Springer, Vienna

Burry M (1999) Paramorph AD profile 139: hypersurface architecture II. Academy Editions, London

Clifford B (2012) Thick funicular: particle-spring systems for variable-depth form-responding compression-only structures. In: ACSA international conference: change architecture education practice, pp 475–481

Dieste E (2004) Architecture and construction: innovation in structural art (Andersons). Princeton Architectural Press, New York, pp 182–190

Evans R (1997) Translations from drawing to building. MIT Press, Cambridge

Kolarevic B, Malkawi AM (2004) Performative architecture: beyond instrumentality. Spoon Press, London

Mitchell M (1996) An introduction to genetic algorithms. MIT Press, Cambridge

Rippmann M, Block P (2011) Digital stereotomy: voussoir geometry for freeform masonry-like vaults informed by structural and fabrication constraints. In: Proceedings of the IABSE-IASS symposium 2011. London

Vogel S (2003) Comparative biomechanics: life's physical world. Princeton University Press, New Jersey

Part II
Projects

Integrated Design and Robotized Prototyping of Abeille's Vaults

Thibault Schwartz and Lucia Mondardini

Abstract This chapter discusses a continuous process allowing the design, simulation, and automation of the robotic fabrication of stereotomic vaults obtained by adapting the stone elements bond typical of a flat vault designed in 1699 by the French engineer Joseph Abeille to a spherical surface. This process is based on the joint use of Wolfram Mathematica for the geometric modelling, and Grasshopper in combination with the HAL Robot Programming and Control plug-in for the feasibility analysis, the automated generation of toolpaths, the simulation of the movements of the robot, the dynamic calibration of the end-effector, and the programming of the machining tasks.

Keywords Robotics · Nexorade · Reciprocal structures · Integrated production · Real-time

The results presented below were obtained during the 5-day workshop on 'Nexorades and Reciprocal Structures' led by Lucia Mondardini and Martina Presepi, with the participation of Thibault Schwartz and Tristan Gobin, organized by Philippe Morel in the Graduate Architectural Design Research Cluster 5 at the UCL Bartlett School of Architecture from 4 to 8 March 2013.

T. Schwartz (✉)
UCL Bartlett School of Architecture, London, UK
e-mail: ts@thibaultschwartz.com

L. Mondardini
ENSA Paris-Malaquais, Laboratoire GSA, Paris, France
e-mail: mondardini.lucia@gmail.com

W. McGee and M. Ponce de Leon (eds.), *Robotic Fabrication in Architecture, Art and Design 2014*, DOI: 10.1007/978-3-319-04663-1_13, 199
© Springer International Publishing Switzerland 2014

1 Introduction

As suggested in Amedée-Francois Frézier's treatise on stereotomy (1737), Abeille's flat vault bond is closely related to the timber frame designed by Villard de
Honnecourt (Thirteenth century) and Sebastiano Serlio (1545): both of them
provide a solution for the problem of covering a space with elements shorter than
the span based on a square tiling. The main issue in applying an Abeille's bond on
any surface is based on a procedure expedient for the design of nexorades, i.e. a
structure made of 'nexors', a beam often having four simple connections, two at its
ends to be supported and two at intermediate points to bear other nexors. The
geometrical construction of this kind of spherical vault (Fig. 1) requires the regular
subdivision of a spherical surface into parts sufficiently similar that they can be
made with the smaller number of different elements. The problem is similar to the
definition of geodes, i.e. a semi-regular polyhedron whose vertices belong to the
same spherical surface, and whose edges lie in planes which contain the centre of
the sphere. As a structure composed of elements of small dimensions that can be
rationalized as families of identical objects, Abeille's vaults seemed to be an
adequate subject for the initiation of students to the issues of automated manufacturing with the following objectives:

- Introduction to the notions of integrated production, rationalization and opti-
  mization applied to the fabrication of structural components
- Analysis of a simple set of design constraints, and their impact on the feasibility
  of a design project
- Data exchange methods between software environments
- Production and assembly planning of a simple structure.

Moreover the use of freestone in contemporary architecture combined with
automatic and computer controlled cutting processes has received renewed
attention by researchers (Fallacara 2006; Rippmann and Block 2011) and industry.
As an Abeille's vault full scale prototype, the production presented in this chapter
is the first one based on a spherical mapping (instead of a planar one) obtained via
6-axis hot wire cutting and assembled without guiding framework.

2 Geometrical Construction

There are several ways to define a geode, but they usually start from a semi-regular
or regular polyhedron. Such a polyhedron is in fact seen as a first approximation of
the spherical shape. Successive approximations are obtained by subdividing the
original faces of the polyhedron into triangles and projecting their vertices on the
surface of the sphere in the direction of its centre. The number of subdivisions into
triangles is called 'frequency' of the resulting geode. The 'Geodesate' function of
Wolfram Mathematica produces the geodesic sphere of wanted frequency starting

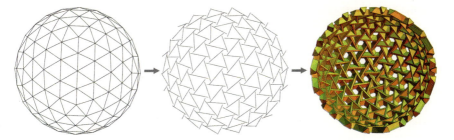

Fig. 1 Model of the hemispherical abeille's vault that was fabricated

from any of the polyhedra primitively known by the code or obtained from them through stellation or truncation. Faces are first triangulated to the wanted order, and their vertices are then projected on the sphere. Starting from all the polyhedrons available in the Mathematica library, the procedure applied to define the geometry (Brocato and Mondardini 2010, Mondardini 2010-2014) allows to generate the geode, to obtain the axis of the corresponding nexorade by using the rotating method (Baverel et al. 2010) and finally to acquire the definition of the quoins shape. The nexors are used as the axis of the quoins so the starting sphere approximately represents the middle plane of the vault. Tests performed with this procedure demonstrate that not all the polyhedrons are suitable for this kind of bond: the more regular they are the easier and more precise will be the final bond.

3 Model Constraints and Export

The geometry of the vault and its components is defined by the following parameters:

- the radius r of the vault sphere, the height b of the ashlars,
- the length a_0 of the mid-segment along an axis that is orthogonal to the nexor and on a plane tangent locally to the sphere,
- the φ splice angle that defines the inclination of the contact surfaces between the ashlars regard to the vertical direction,
- the α chirality angle applied to the axes of the geode in order to obtain the nexors,
- and z_0, used to obtain vaults spanning less than the hemisphere.

Brocato and Mondardini (2011) depicts the influences of these geometrical parameters on the mechanical behaviour of this kind of structures.

The generating polyhedron and the frequency of the sphere are also controlled, allowing to affect the number of quoins and, together with the radius and the final dimensions of the elements, to constrain the size of the structure to make it correspond to the mechanical and material limitations set by the robot kinematics and the EPS blocks format that were chosen to realize this prototype. The algo-rithm developed in Mathematica ends with a rather extended part dedicated to the fabrication of the structure, targeting both rapid prototyping and full-scale reali-sation: for the rapid prototyping, *.stl files are automatically produced, including the shape of the quoin, the trace of the neighbouring quoins to simplify the assembly, and the family which the quoin belongs to. For the full-scale realisation, a script file is streamed to a specified directory, exporting the full geometrical model with a 10^{-6} mm precision for the preparation of the hot wire cutting procedures in Grasshopper using the following data syntax:

$$\% \textbf{\textit{Ashlar}}[\textbf{\textit{Family}} \ \$\textbf{\textit{ID}} \ \$\textit{Geo}[\ast \textbf{\textit{Faces}} \ [/\textbf{\textit{Pts}}[(X;Y;Z)]] \ast \textbf{\textit{Traces}}[/\textbf{\textit{Pts}}[(X;Y;Z)]]]]$$

4 Geometry Reconstruction and Validation

In Grasshopper, a script listening to the Mathematica export directory is dynam-ically loading the file as a string. The data is then parsed to obtain a DataTree of the following structure:

$$\{\textbf{\textit{Family}}, ID\} (Ashlar(as \ a \ BRep), Traces(as \ a \ list \ of \ curves))$$

The export tolerance allows Rhino to recognize the planarity of the ashlars surfaces, which will be used later to generate the corresponding toolpaths. The integrity (number of faces, closed state) of the ashlars is then verified and an analysis of the bounding boxes dimensions is generated for every part, allowing validation for the feasibility of each ashlar. For this fabrication experiment, an ABB IRB120 robot (range of 590 mm, payload of 3 kg) equipped with a 650 mm wide hot-wire cutter was used, constraining the maximum length of the producible ashlars to 600 mm (as the length of active part of the cutter). For aesthetic reasons, the height and width of the maximum bounding box of each part was set to respectively 150 and 200 mm. In order to finalize the geometrical model, an additional section is then performed on the base ashlars so they can connect correctly to the ground. This intersection corresponding to an additional cutting during the manufacturing pro-cess, every part affected by this modification is identified within its family and isolated in a specific sub list of ashlars. The total computing time of the recon-struction, validation and update of the vault geometry ranges from 11 to 12 s, depending on the number of ashlars intersecting the ground plane (computing time measured on a single CPU thread at 3.2 GHz). If one of the various validation feedback was returning a negative result, the students had to modify the different parameters of their vault model until suitable results could be found to fulfil the main technical set of constraints and their own aesthetic choices—such as the density of openings, the asymmetry obtained on the hemisphere, the overall homogeneity of the ashlar sizes, etc. Once a solution was found, they could save it by setting a specific name to the export file. The last selection step—amongst the few dozens of final proposals—favoured the vault with the largest span, also corresponding to a larger value for φ and thus reducing the risks of collisions between the tool and the part support during the cutting process. The final fabricated model was composed of 222 ashlars with the following dimensions: $452 < L < 534$, $H = 150, W = 174$ mm. The obtained hemispherical vault had a span of 3,420 mm for a height of 1,610 mm. Both Mathematica and Grasshopper templates were provided and presented to the students during the first morning of the workshop, which allowed the group to get a valid result at the beginning of the second day, while in the meantime the robot was installed and calibrated using a method that will be depicted later in this chapter.

5 Toolpaths Generation

As each part has to be cut in all directions but has only flat faces, it is possible to place them so that only 5 of their 6 faces (in case of a regular ashlar) are cut by the robot. For this purpose, a pre-cutting of each part was ordered to the foam manufacturer in order to obtain $200 \times 600 \times 150$ mm raw blocks of EPS, corresponding to the maximum dimensions of the ashlars described in the model. In order to ease the quick manual positioning of the blocks in front of the robot during the fabrication, the first position of the toolpath after the initialisation ('home') position (Fig. 2-1) is set to follow the long edge of the block closest to the robot, minimising the rotation error due to the placing of the part. The raw blocks having an acceptable tolerance ($+2$ to $+10$ mm) and being aligned with the wire, they are fixed on one of their $200 * 600$ mm faces, allowing to frontally cut the 3 remaining long faces in one movement (Fig. 2-2 and 2-3). It is then possible to reorient the end-effector above the part (Fig. 2-4) to position the cutter for the lateral engraving and cutting without the risk of collisions (Fig. 2-5).

In addition to the cutting of the faces, the traces of neighbouring ashlars have to be engraved. As these traces are leaning and are oriented on the inclined faces of the part, the engraving can be managed before (Fig. 2-6 and 15) and after (Fig. 2-8 and 18) the cutting of the lateral faces (Fig. 2-7 and 17) as a prolongation of the end-effector lateral reorientation, thus minimising the production time. In the case of an additional cut due to the intersection of the part with the ground once assembled, this movement is inserted after the last tracing of the side where this cut is the most accessible (Fig. 2-11). Once all the cuttings and tracings are performed, the robot goes back to the 'home' position to help with the part extraction (Fig. 2-21 and 22), before looping back to the 'waiting' position where a new block can be placed along the wire. A first implementation of this toolpath as a series of segments in space representing the wire of the cutter can be extracted from the ashlar geometry with the standard Grasshopper tools.

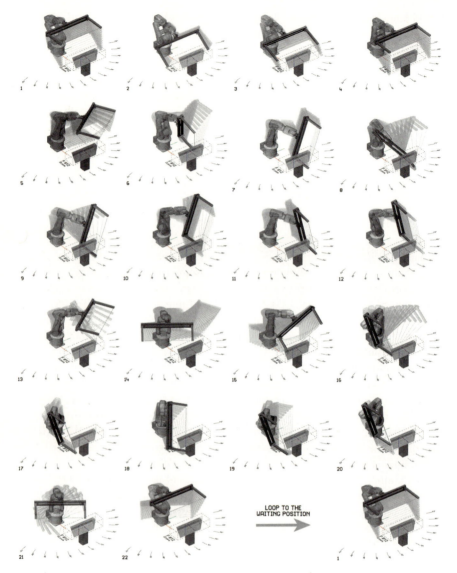

Fig. 2 Steps of the cutting of an ashlar

6 Toolpath Optimisation

Since a hot wire cutter always cuts along its wire, we can optimize the orientation of the cutter by taking two degrees of freedom into account: the first one as a rotation around the wire, the second one as a translation along the wire. These two movements do not theoretically modify the orientation of the cutting, allowing to easily find alternative solutions for a position out of the range of the robot, or causing a collision with the workspace, the robot, the part or the support of the part. The HAL plugin (Schwartz 2012) was used to compute the inverse kinematics of the machine, to detect eventual collisions, and to solve the orientation of the hot wire in a specified domain of rotation around the wire as previously exposed. The translation is used on the lateral cuttings and tracings, helping to only use a portion of the wire temporarily, thus avoiding collisions with the work environment that would occur if the tool centre point was set on the cutting position. The average computing time of the toolpath generation and optimisation procedures, including the inverse kinematics solving and the code generation for all the ashlar families is of 1 s without the collision detection, and 30 s with the collision detection (computing time measured on a single CPU thread at 3.2 GHz).

7 Real-Time Calibration via Smartphone and Upload

The precision of a robotic fabrication process heavily relies—among other things—on the calibration of the end-effector(s) used. The built-in calibration routines available in robot controllers, while being quite adapted to tools being fixed with a simple or null rotation, do not usually provide procedures allowing to interactively refine the calibration data of an end-effector. In order to quickly and efficiently calibrate the hot wire cutter before the beginning of the production phase, an extension of the HAL mobile interface for tablets and smartphones based on TouchOSC (http://hexler.net/software/touchosc), was developed to allow the real-time modification of the geometry of the cutter in the simulation model, thus redefining the tool data declaration taken into account by the robot. In the absence of sensors that would have allowed to automate a position search (i.e. such as proposed by the ABB Rapid *SearchL* command), a verification routine reorienting the tool along its wire/Y direction was looped to verify the precision of the calibration against positioners (pins) placed along the wire. Despite the very short time allocated to this process during the workshop (20 min to record approximately 100 tool positions from which an average tool calibration was deducted), this experimental method proved to be acceptable, providing an orientation precision along the 650 mm wire of less than 0.2°. Once the tool calibration is complete, the production of each family of Ashlars was executed by dynamically uploading the tasks generated to the robot controller from Grasshopper via the ABB PC SDK implementation provided in HAL (Fig. 3). The average cutting time per ashlar at a cutting speed of 17 mm/s was ranging between 150 and 200 s.

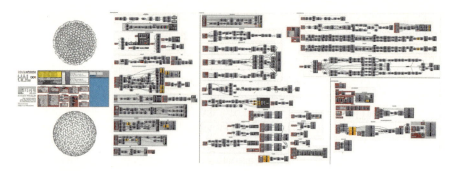

Fig. 3 Production grasshopper definition. *Left* control panel and the import of the Mathematic model. *Middle* toolpath generation. *Bottom right* robot simulation and control

8 Conclusion

The method described in this chapter allowed, in a 5-days period, to set up a hot wire cutting cell involving a 590 mm 6-axis robot, and to produce and assemble a 3.4 m EPS Abeille's vault based on a mixed process covering all the design and fabrication aspects of such a structure. Due to the limited dimensions of the cutter and the manipulator, some solutions of hemispherical Abeille's vaults mapping involving very irregular length of ashlars were discarded but would have theoretically given better structural solutions with a gradually diminishing density of material following the vertical progression of the vault. As a pedagogical experiment, this workshop provided the students a unique and intense experience of production, and a clear view of the relationship established between the different (conceptual and graphical) tools they were learning at this period of the academic year. We also hope that the relatively extreme time, space, and budget limitations applied to this event naturally led them to a better understanding of the elegance that—seemingly—purely technical objects and techniques can produce in highly constrained environments. This process, limited in this example by the characteristics of the machine and the hot wire cutting process allowing to manufacture only foam elements, could eventually be scaled and ported to stone cutting applications using bigger and stiffer machines equipped with diamond saws, as presented by McGee et al. (2012). In this case, we think that this method would allow us to minimize the production time and material loss of structural stone components compared to more invasive material subtraction methods such as 5 or 6 axis milling, as used by Tamborero (2012).

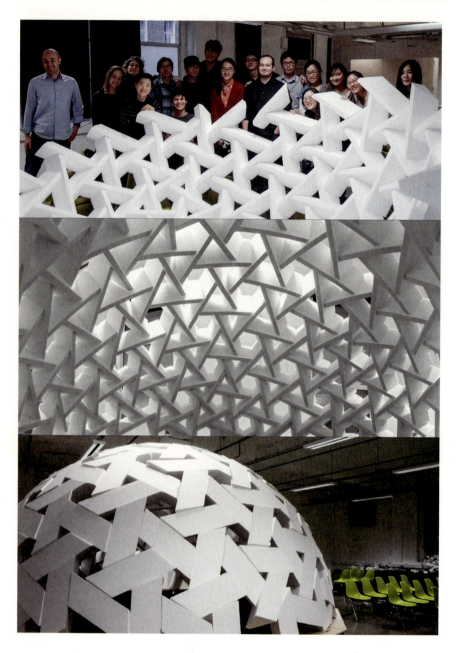

References

Baverel O et al (2010) Nexorades. Int J Space Struct 15:155–159

Brocato M, Mondardini L (2010) Geometric methods and computational mechanics for the design of stone domes based on abeille's bond in advances in architectural geometry. Springer, Vienna, pp 149–162

Brocato M, Mondardini L (2011) A new type of stone dome based on abeille's bond. Int J Solids Struct 49:1786–1801

Fallacara G (2006) Digital stereotomy and topological transformations: reasoning about shape building. In: Dunkeld M (ed) Proceedings of second international congress construction history, Queen's College Cambridge, Cambridge, pp 1075–1092

McGee W, Ferringa J, Sondergaard A (2012) Processes for an architecture of volume: robotic wire cutting. In: Rob|Arch robotic fabrication in architecture, art, and design, Springer Vienna, pp 62–71

Mondardini L (2010–2014) Contribution to the development of stone structures: modelling, optimization and design tools. In: CIFRE 1506/2010, SNBR & ANRT, France

Rippmann M, Block P (2011) Digital stereotomy: voussoir geometry for freeform masonry-like vaults informed by structural and fabrication constraints. In: Proceedings of the IABSE-IASS symposium taller, longer, lighter, London

Schwartz T (2012) HAL: extension of a visual programming language to support teaching and research on robotics applied to construction. In: Rob|Arch robotic fabrication in architecture, art, and design, Springer, Vienna, pp 92–101

Tamborero L (2012) Coupole appareillée à la manière d'Abeille. Al Wajba Spa and Fitness, Doha, Qatar

Mediating Volumetric Thresholds

Detached Fabrication Processes for Non-Uniform Solids

Gabriel Fries-Briggs

Abstract Volume-based design logic has become focal point in fabrication discourse giving rise to the development of new techniques, episteme, and modes of architectural representation. While building on this reconceptualization of material, volumetric design-logic can be advanced by introducing the design and fabrication of non-uniform solids in which volume cannot be described as the matter between surfaces. Detached fabrication is proposed as a method of fabricating both precise and indeterminate internal forms using an industrial robot to mediate the interaction of multiple materials and volumes without direct manipulation of a tool. This method is presented as both a new technique for fabrication with industrial robots and an additional model of designing through volumetric logics.

Keywords Detached tooling · Rapid prototyping · Non-uniform volume

1 Introduction

A recent positioning of volume as a focal point in fabrication discourse has given rise to the development of new techniques, episteme, and modes of architectural representation. Volumetric considerations have exposed architect's over-reliance on surface and the subjugation of design thinking to the constraints of thinness, flatness, and other common denominators of industrially processed materials (McGee et al. 2013). While building on this reconceptualization of material, volumetric design-logic can be advanced by introducing internal deformation of solids in which volume cannot be described as the matter between surfaces.

G. Fries-Briggs (✉)
Princeton University, Princeton, USA
e-mail: gfries@princeton.edu

W. McGee and M. Ponce de Leon (eds.), *Robotic Fabrication in Architecture, Art and Design 2014*, DOI: 10.1007/978-3-319-04663-1_14,
© Springer International Publishing Switzerland 2014

The adoption of 3d printing, 3d scanning, and industrial robotics into the architectural tool-set has generated new conditions for direct volumetric fabrication, but these tools are rarely leveraged to design internal complexity in non-uniform solid forms.

Non-uniform solids are a given condition in natural building materials and industrial material manufacturing has a long history of integrating irregular and unpredictable densities. X-ray scanning logs to maximize yield has been developed in timber manufacturing for several decades (Lindgren 1991). Non-uniform volumes can be not only a given material property, but can be created to obtain a range of material properties such as lightness (aerated concrete) or thermal breaks (spray foam insulation). This chapter proposes a method of fabricating both precise and indeterminate internal forms by using an industrial robot to mediate the interaction of multiple materials and volumes without direct manipulation of a tool. This method is presented as both a new technique for fabrication with industrial robots and an additional model of designing through volumetric logics.

2 Methodology

Detached Fabrication (DF) is a means of internally reorganizing solid volumes to generate new material distributions without slicing, carving, or adding to existing surfaces. The process of DF builds on the potentials of hot wire cutting by replacing direct tool manipulation via the robotic arm with indirect tool control through a combination of gravity-guided free fall and robotic material positioning. Heated steel solids are released at precisely determined temperatures into EPS foam blocks attached to a robotic arm. Scripting software is used to generate the best fit movements of the foam block with the robot and translates them into gravity-based tool paths adjusted for material heating and cooling rates. The phase change that results from melting foam creates hardened internal molds that can be cast (Fig. 1), purposed as single piece rotomolds (Fig. 2), used as pneumatic conduits, produce new structural properties within the foam, and serve as complex pathways for tension systems.

Two of these projects have been explored: one, the creation of internal structural pathways for post-tensioning foam blocks for the rapid assembly of towers (Fig. 3) and two, casting internal networks of structural forms.

3 Case Study: Rapid Prototyping Structural Forms

The primary benefits of creating structural formwork using a detached tooling method are speed, no orbital tool constraints or object conflicts, cost, and no subsequent assembly of components. Single-piece formwork allows complex forms to be cast as single units. Tool entry-points, the only void exposed to the

Fig. 1 Plaster cast of
internal pathways after
removing foam

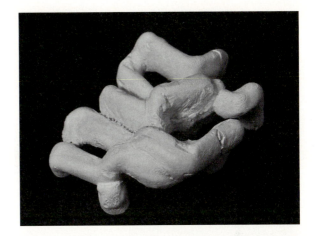

Fig. 2 Foam mold created
for rotomolding

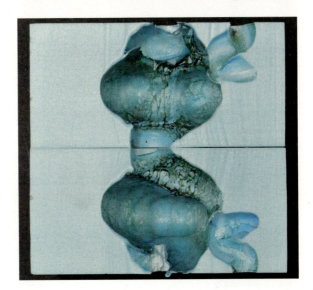

foam's surface, function to deliver plaster or other casting materials to the internal
network of voids. The hardened surface that results from the melting foam allows
the formwork to be manually cut and broken away or rapidly dissolved in acetone
in fragile locations. The range of structural forms possible with detached tooling
greatly expands on those available via milling or hot-wire cutting and is only
constrained in that the volume of formwork must be continuous at some point.

Figure 1 illustrates a form composed of linear pathways where voids larger than
the diameter of the heated steel are calculated either as the intersection of multiple
pathways or produced by rotating the foam perpendicular to the path to enlarge the
melted diameter. The pathways were generated via scripting software to approximate the overall desired form with the selected detached tool (in this case, a

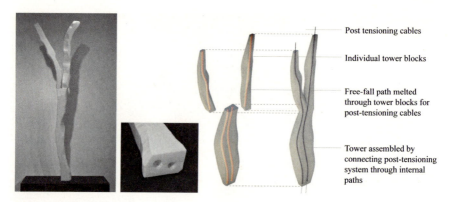

Fig. 3 Photographs of assembled model and diagram of basic tower assembly units with post tensioning pathways

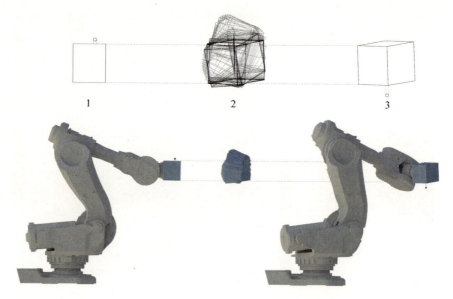

Fig. 4 *1* Robotic arm starting position at ball release, *2* Sequence of rotations performed on the foam block for one pathways, *3* Ending position and exit point of steel sphere

0.75 inch diameter steel sphere). Calculating the inverse kinematics of the specific industrial robot yielded the rotations of the foam block such that the path of the steel maintains a gravity-guided free-fall adjusted for the rate of cooling, often requiring extra rotations about the x-axis to prevent joint limitations (Fig. 4).

The heated steel is delivered to the foam from a fixed location such that the entry point is calculated in the robotic arm's coordinate system. The steel is heated to above 425 °F, placed in the delivery device, and when an infrared sensor measures 425° the ball is released, initiating the rotation sequence (Fig. 5).

Fig. 5 The detached tool release device with infrared thermometer and robotic arm in starting position

Fig. 6 Rotation of foam block in detached fabrication sequence

A 0.75 inch heated steel sphere continues to melt through standard EPS for two-minutes before cooling to the point of sticking before which point it is ejected back through the same path, a previously created void, or creates a new exit/entry aperture in the surface (Fig. 6).

While internal voids formed by a solid steel sphere were created with a relatively high level of precision and predictable output, other shapes and materials were also heated and inserted into foam blocks. Objects of similar mass heated to the same temperature and subjected to the same rotations yielded unique and unreproducible forms. Unlike the forms generated from spheres, other geometric forms (cubes, cylinders, cones, etc.) were each capable of producing a range of formal variations depending on their rotation relative to the path of gravity. Since the resultant form could not be precisely determined, a method of observation and interaction with the material processes was developed.

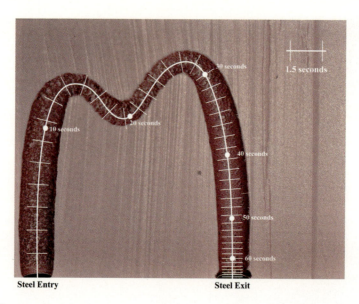

Fig. 7 Section cut through detached toolpath. The narrowing of the path denotes lateral movement of the steel sphere

4 Observation and Representation

DF relies on the development of new representational strategies to accommodate non-solid volumes and fabrication processes that are not visible to the human eye. These strategies take into account a historical perspective that designing non-solid volumes has been tied to the development of new representational techniques. The axonometric drawing became a standard in designing mining operations beginning in nineteenth century England due to its capacity for scientific precision and to make complex, non-uniform masses legible. As fabrication processes have evolved in recent decades to embrace instability and unpredictable materials (Dierichs and Menges 2012), drawings serve not only as tools for design but for measuring and intervening during fabrication. In detached fabrication, axonometric design space is a tool for both designing the interaction of volumes and for representing the real-time material processes.

With the recent emergence of attention to volumetric design protocols in architecture, tools that were once the provenance of climatic engineers and physicists can be used to question what constitutes a physical boundary (Lally 2008) and how energy transfers through materials. DF modeling requires a high level of precision in physical simulations to calculate the falling of heated objects, their intersection with previously melted forms, and their rate of cooling. To achieve reciprocity between the simulation and the fabrication outputs, a series of tests with variable rates of cooling were conducted and measured against the scripted cooling rates (Fig. 7).

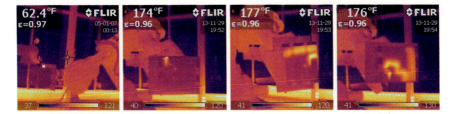

Fig. 8 Thermal images during detached fabrication from two camera perspectives

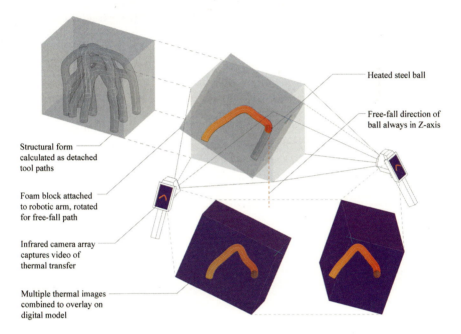

Fig. 9 Diagram of thermal imaging array and digital reconstruction of pathway

A section cut through the foam reveals the final result after several iterations of tests created reproducible accuracy.

Although accuracy can be achieved with spherical tools and materials of consistent density, allowing the stochastic material processes to participate in the generation of DF necessitates forms of observation that deal with geometries that cannot be seen during fabrication. A thermal imaging array (taking thermal photographs from at least two locations) provides an image of thermal tool paths only visible during the interaction of heated steel and foam (Fig. 8).

These images are overlaid on the digital model, creating observation techniques that allow designers to intervene and adjust for cooling, intersection, and manipulations to the desired geometries (Fig. 9). This strategy has been developed to integrate imaging feedback of the fabrication process with the digital model, ensuring that human interpretation continues to inform design throughout robotic fabrication.

5 Conclusion

By using the robotic arm to direct the interaction of volumes, the detached tool is subject to fewer movement constraints and opens up a field of formal possibilities, showing that complex forms can be created without complex end effectors. Since new tooling objects can be introduced into the process with little setup, detached fabrication enables rapid prototyping with a wide range of potential tools. EPS foam and heated steel were the primary material agents in this project, but expanding the potential of detached tooling techniques will require greater experimentation with material interactions and a rethinking of what can function as a tool. In mediating volumetric interactions, surface is taken to be only a threshold; not a static boundary, but a generative one. Building on the logic of intensive and extensive differences (Reiser and Umemoto 2006), mediating volumetric thresholds proposes a new means of staging interactions between materials and allowing matter and material behavior to inform the processes of robotic fabrication.

Acknowledgments Thanks to Axel Kilian and the Embodied Computation Lab at Princeton University for guidance and support. Robotic pathways were generated using the Mussel plugin for Grasshopper, created by Ryan Luke Johns, who also generously provided advice. Additional thanks to research cohorts Charles Avis, Nicholas Pajerski, Brendan Shea, and YshaiYudekovitz for advice and criticism throughout the project.

References

Dierichs K, Menges A (2012) Aggregate structures: material and machine computation of designed granular structures. Architectural Des 82(2):74–81

Lally S (2008) Walking through walls. In: Abruzzo E, Solomon JD (eds) Dimension, vol 12. 306090 Books, New York, pp 20–25

Lindgren LO (1991) Medical CAT-scanning: X-ray absorption coefficients, CT-numbers and their relation to wood density. Wood Sci Technol 25:341–349

McGee W, Feringo J, Søndergaard A (2013) Processes for an architecture of volume: robotic wire cutting. In: Brell-Cokcan S, Braumann J (eds) Rob/Arch 2012: robotic fabrication in architecture, art and design. Springer, Wien, pp 62–71

Reiser J, Umemoto N (2006) Atlas of novel tectonics. Princeton Architectural Press, New York

Instruction and Instinct

Performing Within Unordered Complexity

Emmanuel Vercruysse, Kate Davies, Tom Svilans and Inigo Dodd

Abstract The chapter addresses the relationship of robotics to complex or unpredictable site conditions. Robotics is generally explored in controlled environments and lab spaces. RAVEN actively experiments with the interaction of robots and natural or external environments and the chapter discusses the testing phase for two robotic performance works, *Pyro* and *Hydro*—using UR10 robots as test rigs for designing choreographed events within spatially complex and environmentally dynamic conditions. It explores the poetics and logistics of robotic field operations and posits an artistic engagement with robotics in relation to landscape or other external environments beyond industrial or military applications. Raven seeks to create performance-based interactions which set up a discourse about place, site, and location. We design choreographies that explore notions of precision and repeatability in conversation with chaotic or unpredictable environments.

Keywords Environment · Interaction · Performance · Choreography · Control

E. Vercruysse (✉) · K. Davies
LiquidFactory and The Bartlett, UCL, London, UK
e-mail: emmanuel.varcuysse@ucl.ac.uk

T. Svilans · I. Dodd
The Bartlett, UCL, London, UK
e-mail: tom.svilans@gmail.com

I. Dodd
e-mail: i.dodd@ucl.ac.uk

W. McGee and M. Ponce de Leon (eds.), *Robotic Fabrication in Architecture, Art and Design 2014*, DOI: 10.1007/978-3-319-04663-1_15,
© Springer International Publishing Switzerland 2014

1 Natural Habitats

The natural habitat of robots is a controlled environment—the lab, the factory, the utilitarian fabrication space—white, unassuming, unspectacular spaces. In order to perform tasks with precision and accuracy, the prerequisite spatial condition of these realms of operation is control.

There are, however, robots outside the lab; in hazardous, extreme or remote landscapes—the preserve of the military and the space program or of the mining industry and agriculture. Robots are used to negotiate contaminated land or minefields; to navigate war zones and impenetrable terrain; for precision agriculture or remote mining operations. Take, for example Boston Dynamics' BigDog or RHex rough-terrain robots (http://www.bostondynamics.com), GPS controlled John Deere harvesters (http://www.forbes.com/sites/jenniferhicks/2012/08/06/intelligent-sensing-agriculture-robots-to-harvest-crops/) or the driverless excavations of Rio Tinto's 'remote control mining' (http://www.popsci.com/technology/article/2013-09/mining-company-uses-robotic-trucks) where mine truck drivers sit in the comfort of a control centre at Perth international airport operating their machines remotely in the Australian Outback. These roving robots require one key common ingredient; live feedback with which to sense their position and relationship to the landscape they occupy. This allows the site or 'field' of operation to be continually calibrated and recalibrated.

Inspired by these 'wilder robots', RAVEN undertakes field operations outside the lab, exploring dynamic robotic interactions within natural or complex systems.

2 RAVEN Robotic Field Operations

RAVEN is a cross-disciplinary research practice (operating alongside the Bartlett School of Architecture's B > MADE fabrication facility). We have a deep interest in the technical aspects of the work but ultimately we look to engage in a wider poetic, spatial, and cultural discourse through the robotic fieldwork we undertake. RAVEN actively experiments with the interaction of robots and external environments—sites not of control or consistency but of change and dynamism—appropriating these and associated technologies to enact site-specific performance works. The experiments presented here are part of a preliminary test phase for developing the choreographed sequences of movement for these performances and the design of the end-effectors, which act as physical, sensorial, and sculptural interfaces with the environment as they perform dynamic movement sequences. We use the precision of robotic motion and the accuracy of imaging and sensing technologies to explore these sites as hybrid digital-analog realms. It is within this domain that we are able to construct a reflexive relationship between landscape, code, and action.

The operations stray from the tightly controlled enclosure of the laboratory into the grit and imprecise territory of fieldwork. This demands consideration of alternate design constraints and an expansion of the logistical infrastructure necessary to transform the site into a suitable setting for the work. This can be divided into two categories: the transformations and translations between multiple frames of reference in order to precisely locate each actor within its relation to the others and the site; and the control elements required to translate sensory input and mechanical constraints into planned trajectories and actions. A set of sensory tools and locating devices such as laser scanners and joint sensors constitute the first, enabling precise analysis of positions and orientations while also documenting the unfolding surface of the new 'field laboratory'. Software, motion planning and real-time feedback handling make up the second bracket. Live sensory input is used to adjust the robot goals based on environmental factors (fire, wind, pressure, proximity) while motion planning and inverse kinematic solving drive the robot towards the resultant transformations.

Counterpoint to both is the inevitable randomness and unpredictability, which infect the work with an exciting complexity and opportunity. Inherent in the move towards 'the field' is the embracing of the complications and disruptions this move entails, as well as the tentative release of imposed logic and control. In a similar way, the choreographic exploration of the robot as a set of mechanical limits and rotational movements seeks to undermine the typical requirement for the robot to move in prescribed lines or arcs, and instead to play with the more complex—but inherent—motions and relations between joints and components outside of the Cartesian space of the lab.

3 Pattern Practice

In martial arts, Pattern Practice or in Japanese, Kata (literally 'Form')—which is also used in theatre and ceremony—is the repetitive rehearsal of a set of movements in order to program the body through muscle memory to perform them instinctively (Friday and Seki 1997). The Kata are not thought of as rigid but rather as a set of reflexes to be executed without thinking in combat. They are combined without hesitation into reconfigured combinations that adapt to multiple circumstances. In the fight there exists a beautiful fusion of precision, instruction and programming with instinct, and adaptability. We are similarly interested in the relationship between pre-programmed movements and those that come about as a result of a response to a shifting environmental condition—where sets of defined movement sequences are modulated or adjusted by sensor input. In this way the purpose designed robot attachments act as a dynamic interface between robot and environment, between technology, tool, and nature.

Outlined below are the logistics and specifics of each of the performances and their test methods and sites. Through these tests we developed a series of discrete movement sequences or phrases that can be performed in any order. Through a grammar comprised of primary movement sequences and linking segments that serve to connect the primary movement sequences to each other in space. These movement sequences can be modulated by the sensor input, creating multiple combinations and compositions.

4 Pyro

Fire flares and light radiates from a centre—in a negotiation of the woven forest space, a site of unordered complexity, the reflector responds to the changing intensity of the fire, the young tentative flames, the raging burn, the gentle final embers.

This work explores the negotiation of a spatially complex forest environment, as well as using fire as a dynamic, unpredictable element to which the programmed movements respond. In this way the robot negotiates precision and adaptation throughout the duration of the performance. The Faro 3d scanner was used to capture the forest, laser scanning the trees surrounding the robot site giving precise point-cloud data of the complex spatial environment, turning an uncontrolled environment into a fully calibrated site for the robot, one where the movement sequence of the UR could be defined digitally in precise relationship to the surrounding trees. From this we defined an initial choreography that enacted what we refer to as a 'calibration of the space' by the robot.

The end effector element is a polished mirror surface. It constructs two kinds of reflection, one (digital) constructing a ghost space behind the mirror in the 3d scan, (Fig. 1) and the second (optical) by reflecting the light of the fire directly at the camera recording the event (Fig. 2). In rehearsal for the choreography a series of scanning tests were undertaken, scanning the mirror attachment as it performed pre-programmed movements, to explore the form of the 'ghost space' constructed by the reflection of the laser in the mirror. The Robot was then deployed in the forest site and the various choreographed movements run in sequence, while simultaneously being 3d scanned and filmed.

4.1 Hydro

The water swills and the still, flat Loch is disrupted. Sculptural forms interact with the turbulence. They are agents of interference, grappling with ebbs and flows of their own creation.

Designed for a specially equipped vessel to be sited on Loch Lomond, the performance is inspired by the dexterous paddling action of master oarsmen and the subtle relationship of this basic but elegant tool to its liquid site. It requires the design of a sculptural fin attached to a UR10 robot that is manipulated in a defined sequence to push and pull in the water, performing a floating choreography across the surface of the Loch.

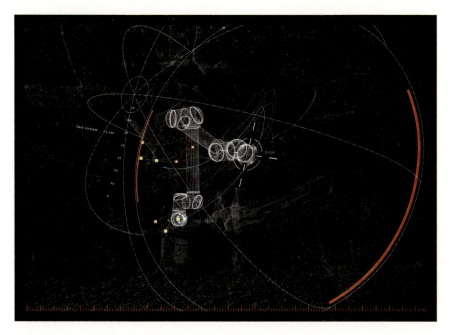

Fig. 1 Simulation of Inverse kinematic tool path inserted into 3d laser scan of forest

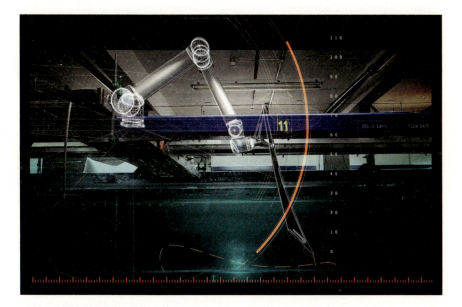

Fig. 2 Movement simulation and force feedback overlaid on actual footage

A series of digitally modeled end-effectors were initially evaluated in in *RealFlow* fluid simulation software to test relationship of form to flow. UCL's Naval Engineering flow tanks, were the site for a physical test phase for exploring and developing the choreography sequence alongside the geometry of the attachments. Using the force-feedback on the UR10 we generated readings of the force exerted by the water, in order to set up a responsive relationship between the motion of the robot and the force exerted on the end-effector.

5 The Poetics of Robotics

By taking the robots out of the lab to construct performance-based interactions on site, RAVEN seeks to provoke discourse about place, site and location in relation to robotics. The design both of the attachment objects and the choreography sequences of these attachments has been developed through a series of iterative testing, fieldwork and importantly, rehearsal. Through these rehearsals, in a way that is perhaps more akin to theatre or dance practice, we are in search of serendipitous moments of fusion between tool and site. These choreographies situate the precision and repeatability inherent in robotics in relation to chaotic or unpredictable environments and elemental factors, defining gestural and irrational motions not born from efficiency and suggestive of the intangible and the metaphysical, the ineffable and the sublime. Choreographer Wayne McGregor describes creativity as 'misbehaving beautifully' (http://www.ted.com/talks/wayne_mcgregor_a_choreographer_s_creative_process_in_real_time.html) a phrase which captures our attitude towards robotics. We are not concerned with direct applications for industry, but rather with the inherent possibilities of beauty and power, theatre and magic, and ultimately with how robotics in architecture, art, and design can be seen as a powerful instrument for creative speculation.

Reference

Friday KF, Seki F (1997) Legacies of the sword: the Kashima-Shinryū and samurai martial culture. University of Hawaii Press, Honolulu

Objects of Rotation

An Opportunity-Driven Design Research Project Integrating the Traditional Craft Based Tools and Methods of Ceramic Wheel Throwing with the Digital Control and Automation Techniques of Robotics

Rachel Dickey, Jili Huang and Saurabh Mhatre

Abstract This chapter describes the intent and goals for the Objects of Rotation project which brings together the contradictory qualities of perceived digital control with those of a process that has traditionally been dependent on human touch, sensitivity, and innate responsiveness to material production and craft. The project pursues an opportunistic approach to design research by exploring the potential of ceramics as a malleable material and by integrating ceramic traditions and tools with automated strategies and robotic technology.

Keywords Robotics · Digital fabrication · Ceramics · Throwing · Pottery wheel · Automation

1 Introduction

The story of our relationship with clay is the story of material culture. It is the story of domesticity, and the story of technological advances. The inventions of the wheel and the kiln, the understanding that fire could turn mud to stone, were the foundation of the thousands of technologies that have followed (Staubach 2005).

R. Dickey (✉) · J. Huang · S. Mhatre
Harvard University Graduate School of Design, Cambridge, MA, USA
e-mail: rdickey@gsd.harvard.edu

J. Huang
e-mail: jhuang@gsd.harvard.edu

S. Mhatre
e-mail: smhatre@gsd.harvard.edu

W. McGee and M. Ponce de Leon (eds.), *Robotic Fabrication in Architecture, Art and Design 2014*, DOI: 10.1007/978-3-319-04663-1_16,
© Springer International Publishing Switzerland 2014

Ceramics have played a crucial role in the history of innovation and in the evolution of human progress. Understanding this role in relationship to its place in the home with vessels and stoves, in the development of machinery and industry through the production of the kiln and potter's wheel, and its role in architecture from the primitive hut to the curvilinear forms of Antonio Gaudi, the Objects of Rotation project strives to continue this evolution by providing new design opportunities and digital tools for the production of ceramics. The project brings together the contradictory qualities of perceived digital control with those of a process that has traditionally been dependent of human touch, sensitivity, and innate responsiveness to material production and craft. Through the integration of the ancient craft of throwing with the precision and control of robotic technology, the project seeks to not only establish a connection between craft based processes and digital fabrication techniques, but also attempts to provide design opportunities through the development of new production tools and automation strategies.

2 Ceramics and Wheel Throwing

A review of the role of clay in the progress of civilization, an explanation of the process of making of wheel thrown vessels, and a description of the role of ceramics in architecture provide a theoretical foundation for this research. In the book, Clay: The History and Evolution of Humankind's Relationship with Earth's Most Primal Element, Suzanne Staubach describes how the making of clay vessels led to the development of the first machines and of an industry and particularly how the potter's wheel was among the earliest of technological inventions used for ease and speed of production. Drawing from this notion of clay and the potter's wheel being crucial to human progress, the Objects of Rotation project combines this ancient craft with new technologies not only to demonstrate new design potentials, but also to conceptually demonstrate a method of research grounded in looking to the past as a way to move forward.

The skilled and poetic process of watching, engaging in, and understanding how a potter throws a vessel on a wheel was fundamental for the research team to understand how clay responds to touch and tools while on a wheel. Staubach describes the wheel throwing process when she writes:

> Watching a skilled potter throw on the wheel is mesmerizing […] With the wheel in motion, the potter dribbles water on the spinning lump of clay and places both hands lightly around it. The clay spins between the potter's hands, the centrifugal forces pushing outward as the potter's palm or palms push in ward. Within minutes, the clay becomes centered, a smooth cone of clay so symmetrical that is motion is almost imperceptible (2005).

As the diagrams illustrate in Fig. 1 throwing a vessel is largely dependent on the amount of force applied to the clay, wheel speed, and the displacement of the clay throughout the making process.

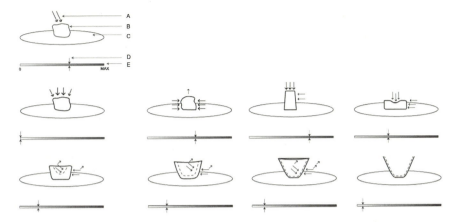

Fig. 1 Process of making traditional wheel thrown vessels. Diagram labels: *A* direction and amount of pressure, *B* clay, *C* pottery wheel, *D* speed marker, *E* wheel speed gauge (from *left* to *right* minimum speed to maximum speed)

Ceramics have also played an integral role in architecture. Clay in some form has given us "shelter from the simplest mud and sticks of huts of the early Stone Age to the multistoried apartments that 'scrape' the skies of modern urban centers" (Staubach 2005). With its often abundance and locality along with its thermal dynamic qualities, there are many reasons why the material has proven useful as a material of construction. However, despite clay's malleable properties, the brick which is rectilinear and standardized is the most common ceramic material for construction (Staubach 2005). The Objects of Rotation project seeks to break away from these traditions and provide new tools which embrace clay's intrinsic material properties.

3 Automation Strategies and Fabrication Processes

A review and understanding of common fabrication processes is essential to understand the relativity and position of the Objects of Rotation project within the realm of design research. The common categories of tools and manufacturing process include: Subtractive Processes or Material Removal; Molding, Deformation, and Casting; and Fabrication or Additive Processes (Schodek and Bechthold 2005). While the project draws from some of these existing methods of manufacturing such as subtractive processes, it also has the potential to add an alternative category which involves shaping or forming. The integration of the wheel in the manufacturing process contributes to the production of alternative or asymmetrical objects of rotation.

Consideration of material properties is essential to any discussion of manufacturing processes (Schodek and Bechthold 2005). More specifically with the use of ceramics, the properties of malleability, change in soft and hard states, and shrinkage after firing are all relative to what kinds of production applications are possible. The previous tools and ways of making in ceramics are core to the material's qualities thus they have the characteristics to inform the fabrication processes of the digital. In order to achieve desired effects and design intentions, new digital manufacturing techniques are incorporating notions of craft and emphasizing the role of material investigation (Kolarevic and Klinger 2008). Similarly, the Objects of Rotation project develops digital techniques inspired by craft and materiality, but more specifically looks to traditional methods and material considerations related to ceramics in order to provide designers with opportunistic tools and processes.

4 Research Problem, Hypothesis, and Questions

With the research team's base understanding of the traditional methods of making ceramic objects and their overview of the strategies in digital fabrication, the realization of conflict arose from the lack of existing connections between the two processes. Thus, the intent of the Objects of Rotation project is to explore the potential interrelationships between digital fabrication and traditional hand-based techniques and to introduce a new way to process ceramic materials into the medium that coordinates the precision of digital production with its inherent material properties and predicable reactions. Through this description of the research problem and investigation of both processes, the design team hypothesized that it is possible to develop a level of control over the surface-treatment and delineation of thrown clay vessels that cannot be matched by traditional hand-based techniques and that the linking of the two methods allows for more opportunities in ceramic digital fabrication.

The opportunities for ceramic fabrication provided by this tooling include the ability to work at a variety of scales with ceramic objects while also providing an increase in precision and a larger degree of replicability. While traditionally the production of clay objects on the wheel is limited to a scale which is feasible for the potter to control by hand, the integration of digital automation allows for these techniques and methods to be scaled in both directions, to allow for more control in the production of both smaller and larger objects. Also, the development of this tooling increases the ability to replicate the objects produced by having a defined system and digitally controlled process.

The project does seek to obtain a certain degree of precision; however, it doesn't seek the same precision of that achieved by the human hand, but instead an alternative precision achieved through the control of integrated tooling and robotics in order to develop a digital craft. To further articulate the concept, the project seeks to provide an opportunistic approach to design research by integrating ceramic traditions and tools with the control of robotic technology; however, it does not strive to provide a robotic mastery of the skill of wheel thrown vessels. It instead combines the knowledge of the craft with the mastery of control from robotics in order to provide an additional tool in the ceramic making process. The evaluation criteria and essential questions for the project include the following:

1. Can digital and hand-craft techniques be merged to reveal new design opportunities? Where do the opportunities for the combination of digital and traditional reveal themselves? What are the advantages/disadvantages of the application of digital fabrication methods applied to hand-craft techniques?
2. Can the whole process be defined in order to produce a specific outcome?
3. How can the project provide a way for the users/designers to understand to the tooling language in order to predict the outcomes for their designed shapes?
4. What makes this exercise potentially relevant to the larger (design) community? Is there a design goal to surpass that which is achievable by human and wheel? What are the opportunities in digital production for this particular process?

5 Research Methods

The design goal for the project is to surpass that which is solely achievable by the human hand with the control and accuracy of robotics. While precision is a term commonly used in the discussion of digital fabrication projects, the precision of this project does not attempt to rely on the same precision and intuition of the potter, but does, however, involve accuracy in the speed of the wheel relative to the placement and timing of the robot. In order to determine the relationship between the robot speed and the rotational speed of the wheel the design team developed a series of calibration prototypes that involved syncing the speed of the wheel and relationally equating the speed of the robotic arm. The team produced a series of basic equations to synchronize the speed of the wheel with the speed of the robot (as shown in the equations Sect. 8). The independent (constant) variables in the research include: consistent diameter clay cylinders (sized at 9 inches tall by 4 inches in diameter), the location of the cylinders centered on the wheel, and the wheel speed. Figure 2 illustrates the setup and components utilized by the design team to test these ideas.

The dependent variables in the research include the tool shape and the tool path which involves the depth of the cut and the tool location in relation to the initial cylindrical clay object. The use of various tool sizes and shapes allow for the production and exploration of how they affect the clay medium in different ways while the varying tool path allows for freedom of movement of the tool in concert with the rotation of the cylinder. The tools utilized are traditional ceramic carving tools that are attached to the robotic arm via a collet chuck. Some of the tools and their clay carved results are illustrated in Fig. 3.

The design team set up a series of challenges to test the level of control in relation to wheel rotation, robot timing and placement. Some of these challenges

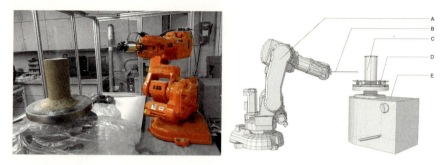

Fig. 2 Research setup and components. Diagram labels: *A* ABB-IRB 140 Robot Arm, *B* versatile tool pack, *C* ceramic cylinder, *D* mounting fixtures and centering device, *E* pottery wheel

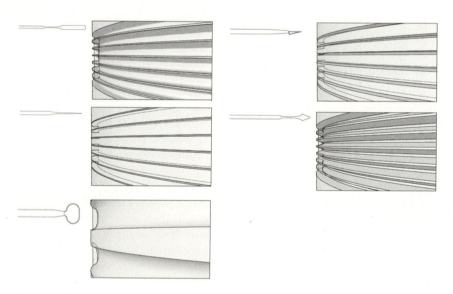

Fig. 3 Traditional ceramic tools and carving outcomes

included carving a straight line on the cylinder and carving a helix shape starting from the base of the cylinder and ending at the top. These challenges were essential to development of the control necessary to make a predictive process. In relation to these challenges a series of objectives the design team set also include: developing a set of proofs that begin to reveal the possibilities of the interrelation between the digital (robotic movement) and the traditional (potting wheel) approaches to processing ceramic material focusing on precision and reliability of the process, describing the process along with the successes and failures, and designing an interface that allows other outside users to design and produce results with this technique.

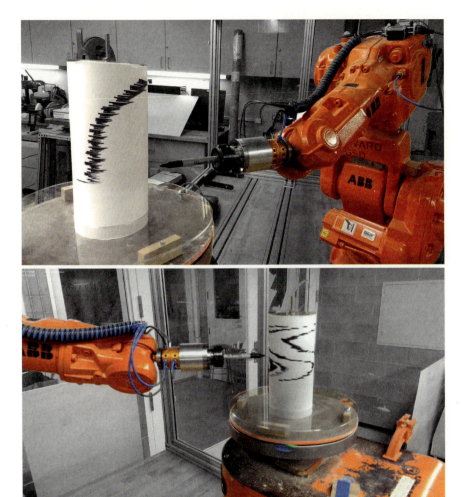

Fig. 4 Research tests with rotating paper cylinder and marker

The development of this ceramic tooling technique was not just about the production of the tool, but was also a way of researching the algorithm and understanding the material responses to different tooling and tool paths. The three outputs the design team produced to understand the results of the testing included predictions, paper tests, and clay tests. In order to for initial testing of control and calibration, the research team used paper wrapped around cylinders and a marking tool (Fig. 4). After achieving the desired results from the paper tests the design team moved forward testing clay cylinders. Thus, the research tests resulted in three versions for comparison including predictions, paper cylinder calibration tests, and resulting carved clay objects (illustrated in Figs. 5, 6, 7).

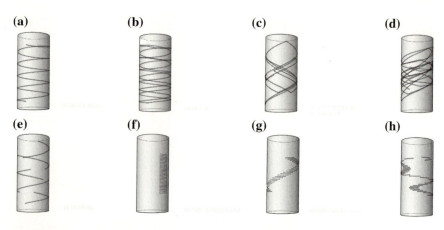

Fig. 5 Research test predictions. **a** single helix, **b** double helix, **c** continuous net, **d** continuous net alternate 1, **e** interval, **f** straight line, **g** dashed single helix, **h** segmented helix

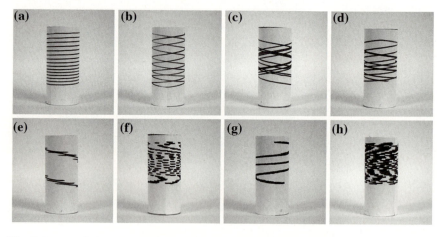

Fig. 6 Paper cylinder tests. **a** single helix, **b** double helix, **c** continuous net, **d** continuous net alternate 1, **e** interval, **f** pattern, **g** dashed single helix, **h** dense pattern

6 Research Evaluation

Based on the research predictions and tests the research team was able to draw a series of conclusions about the degrees of success established in the testing process and gain an understanding for how the project could be developed further. The predictions set forth a place to begin the testing process with a goal for certain achievable results. The paper tests allowed for calibration of the robotic arm with

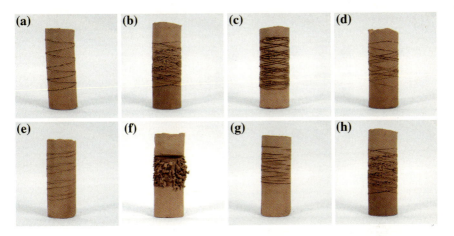

Fig. 7 Clay cylinder tests. **a** single helix, **b** double helix, **c** continuous net, **d** continuous net alternate 1, **e** interval, **f** straight line, **g** dashed single helix, **h** segmented helix

the wheel and testing of the algorithm to evaluate correct placement of the tool in the relation to the rotating cylinder. This system of predicting and testing proved useful for means of comparison. However, with the testing of the clay cylinders the research team became aware of other issues relative to the how the clay responds to various kinds of carving tools and how that relates to the tool path and closeness of the initial carving stroke from the next (similar to the "step over" of a tool path in CNC routing devices). Therefore, while the tool allows for a certain degree of prediction relative to the material's response to it, the objects produced are not always precisely replicable, although similar effects are achievable because the process is infinitely repeatable.

With the current setup of the potter's wheel and the robot, monitoring the exact speed of the wheel was problematic; however, if used in an industrial setting this monitoring of speed could be easily solved with the use of an external axis robot controlling the rotation speed. With this further development of the control in tooling the opportunities for this research could expand to include both shaping and carving capabilities. Shaping with the digital control of the robot and the digital control of the wheel rotation could evolve to the production of objects of rotation and have the potential to develop into a multi-layered process with a series of steps involving an initial clay shape which is shaped to a desired geometry and then textured with the final carving pass. Other options could also include three-dimensionally scanning a clay object and or selecting from a predefined set of shapes. Figure 8 illustrates these options along with the interface and production process.

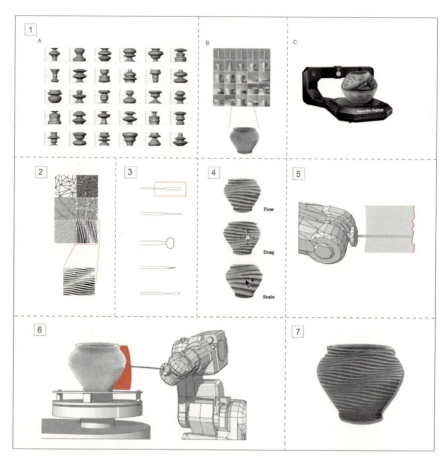

Fig. 8 Production and tooling process. Diagram labels: *1A* select and input shaping strategy, *1B* select predefined geometry, *1C* 3d scan clay object, *2* input pattern to be carved, *3* select tool type, *4* map pattern to geometry, *5* simulate tool path, *6* carve clay object, *7* resulting carved clay object

Fig. 9 Diagrams illustrating change in tooling position

7 Conclusions

Precision today is often so deeply associated with technology it is easy to forget that it is a human ambition first and a mechanic directive second. The logic of precision with its exacting nature does not necessarily increase in coherence with the implementation of digital technology, rather it becomes simply the clearest and the cleanest pathway for the delivery of ideas, whether or not they first emerge from precise thought (Marble 2012).

While the research provides a new tool and technique for producing clay objects with a level of control previously unable to be achieved, it is just a start to the development of tooling and fabrication strategies that could be further developed to have additional architectural implications. The project interfaces where the hand would traditionally be used, not to eliminate it from the making process, but to provide new fabrication opportunities and perhaps foster new ideas about digital craft in ceramic making processes. The project relies on the designer to explore the opportunities the tool can provide and to raise questions and suggestions for how to develop it further. With this role the designer is able to explore the tool's potential for further innovation and actualized creativity. The research team has armed the designer with the tool, an interface to use it, and examples from the production process; now, the designer has the pivotal position to engage in furthering the progress and explore its potential.

8 Equations

Wheel Speed (60 revolutions/second) and Robot Speed (millimeters/second)
   Variables:

t	time (seconds)
X	position of the tool
n	first node (initial position of the tool)
$n + 1$	second node (second position of the tool)
$X_{n+1} - X_n$	change in position of the tool along the cylinder
V	velocity of the robot (millimeters/second)
r	radius of cylinder (millimeters)
d	distance of the tool from the cylinder (millimeters)

Figure 9 illustrates the distance of the tool from the cylinder and how the distance varies when the distance of tool is changing position in the z-axis.
$\frac{2\pi r}{1(s)}$ speed of the wheel (revolutions/second)

Equations for when the tool path is in the same position:

$$\frac{2\pi r}{1(s)} = \frac{X_{n+1} - X_n}{t(s)} \qquad t = \frac{X_{n+1} - X_n}{2r} \qquad d = \frac{V\left(\frac{mm}{s}\right)t}{2r}$$

Equations for when the tool path is changing position:
Wheel Speed (60 revolutions/second) and Robot Speed (millimeters/second)

$$\frac{2\pi r}{1(s)} = \frac{2\pi r - X_{n+1} + X_n}{t(s)} \qquad t = \frac{2\pi r - X_{n+1} + X_n}{2\pi r} \qquad d = \frac{V\left(\frac{mm}{s}\right)t}{2} - 2$$

Acknowledgments This research was conducted under the guidance of instructors Nathan King and Rachel Vroman during the course: Material Systems: Digital Design, Fabrication, and Research Methods at the Harvard University Graduate School of Design; Cambridge MA; Fall 2013. Research supported by ASCER Tile of Spain; Harvard University Graduate School of Design, Design Robotics Group; the Office for the Arts at Harvard, Ceramics Program; and the Harvard Graduate School of Design, Fabrication Laboratory. The software used to conduct the research for this project included: Hal Robotic Programming and Control, Robot Studio, and Rhinoceros.

References

Kolarevic B, Klinger K (2008) Manufacturing material effect rethinking design and making in architecture. Routledge, New York, pp 5–24

Marble S (2012) Digital workflows in architecture: designing design—designing assembly—designing industry. Birkhäuser, Basel

Schodek D, Bechthold M (2005) Design and manufacturing: CAD/CAM applications in architecture and design. Wiley, Hoboken

Staubach S (2005) Clay: the history and evolution of humankind's relationship with earths most primal element. Penguin Group, New York

D-FORM

Exploring the Combination of Laser Cutting and Robotic Thermoforming as a Technique for Architectural Envelopes

Renate Weissenböck

Abstract The aim of this study is to explore the potential of robotic thermo-forming techniques to develop procedures for the construction of facade elements, surface treatments and apertures. A novel fabrication process is created by combining robotic fabrication with laser cutting, as well as by implementing robotic thermoforming techniques. In this exploration, a series of methods, techniques and operations are conducted producing several customized prototypes with individual shaped surfaces. By using digital parametric design tools Rhino, Grasshopper and HAL, adjustments in the fabrication process can be made easily and a variety of elements can be created with similar procedures for varying demands.

Keywords Robotic fabrication · Digital fabrication · Laser cutting · Thermoforming · Parametric design

1 Introduction

In recent years, the accessibility of digital design and fabrication tools provides manifold possibilities for applications in architecture. Industrial robotic arms are intensely explored by architects and designers, as seen at the first Rob|Arch conference in 2012, "...re-using industrial robots as a well established basis and adapting them for architectural purposes by developing custom software interfaces and end-effectors" (Brell-Cokcan and Braumann 2012).

Due to the availability of new interfaces and software, architects can access robots easily and program their processes parametrically. Because of this potential,

R. Weissenböck (✉)
Graz University of Technology, Graz, Austria
e-mail: weissenboeck@tugraz.at

W. McGee and M. Ponce de Leon (eds.), *Robotic Fabrication in Architecture, Art and Design 2014*, DOI: 10.1007/978-3-319-04663-1_17,
© Springer International Publishing Switzerland 2014

it is indicated to look at robotic fabrication in response to performative issues in which elements are optimized for different requirements based on specific site conditions or programmatic demands. In the current discourse of architecture, environmental issues become more and more important; a lot is researched on "...Performance-Oriented Architecture as an integrated approach to architectural design, the built environment and questions of sustainability" (Hensel 2013). To achieve the best performance, repetitive building elements have to be designed and built in variations responding to local needs.

The aim of this research is to explore the potential of robotic thermoforming techniques to develop procedures for the construction of facade elements, surface treatments and apertures to generate a variety of elements with similar procedures for different demands (Fig. 1). A novel fabrication process is created by combining robotic fabrication with laser cutting, as well as by implementing robotic thermoforming techniques, using an industrial six-axis robotic arm (Fig. 2).

This study is placing the technique of thermoforming in the context of architecture and combining it with robotics. Heat deformation has traditionally been used in industrial design, i.e. for packaging, in repetitive processes. By using the digital parametric design tools Rhino, Grasshopper and HAL to generate the robotic procedures and codes, adjustments in the fabrication process can be made easily. The applied deformation is neither additive nor subtractive, but "deformative" by stretching the material.

In this exploration, a series of methods, techniques and operations are conducted producing several customized prototypes with individual shaped surfaces. The used fabrication tools are an Epilog Laser cutter and an ABB IRB 140 robot - a 6-axis industrial robotic arm - combined with hot and cold air.

An experimental approach is used to explore the potential of the developed technique towards adaptive and dynamic forming for architectural applications, adding additional relevance to the use of industrial robots in architecture.

2 Related Works, Extension and Innovation

This research is based on previous experiments with robotic thermoforming using industrial robots. Related precedents are prototypes produced at University of Innsbruck's REX|LAB in the introductory workshop (http://vimeo.com/64710189) as well as in the seminar IsoPrototyping, where the Robo[pusher] was developed (http://vimeo.com/69633175). Also related are explorations done at workshops by the Association of Robots in Architecture at IUAV in Venice (http://vimeo.com/6857276) and at the CAAD Futures 2013 conference in Shanghai.

In this study, this line of research is continued and extended. The laser cutting process is applied to the material prior to the thermoforming process, creating slots, openings or textures. The forming process is conducted by a robotic arm using different robot paths and different shapes of deformers. Because of cutting and scoring of the surfaces, the panels deform differently and create or change

Fig. 1 Prototypes showing different shapes, transparencies, apertures, surface treatments

Fig. 2 Robotic set-up and production process of prototype A deformed by a sphere

apertures due to their deformation. Diverse perforations, patterns, textures and transparencies can be applied by the laser cutter.

The goals and processes defined in this research are directed towards architectural applications for building envelops and inspired by precedents in architecture, where repetitive elements are used with local variations.

At the Elbphilharmonie in Hamburg, by Herzog and de Meuron, similar facade panels are generated according to different aesthetic and performative needs. The repetition of the glass panels in different shapes and transparencies creates a uniform, yet diverse appearance.

The AIA Pavilion in New Orleans, by Gernot Riether, shows the variation of one similar module. "The 300 modules of the pavilion were all different but part of the same family. Each was a different size and proportion, but shared the same base geometry of triangles" (Riether and Jolly 2011).

3 Set-Up: Elements and Processes

After laser cutting, the panels are placed into a custom-made wood frame that is attached to the robot's flange. The deformation of the flat surfaces is created by the robotic arm, moving the frame together with the panel along a pre-defined path and pushing it against a counterpart called "deformer". Depending on the point and depth of deformation in relation to the laser-cut pattern, different sizes and shapes of apertures are created.

To enable flexible fabrication of the prototypes, the robotic path is set-up parametrically. By using digital parametric design tools, quick adjustments in the fabrication process can be made easily according to different requirements. Software used include Rhinoceros 3D and Grasshopper, its integrated graphical algorithm editor, combined with HAL, a Grasshopper plug-in for industrial robots programming, to generate the robotic procedures and the RAPID code (Fig. 3).

In this explorations, the fabricated panels are shaped according to the following "influence factors": material properties, material thickness, shape and size of frame and panel, cut and score patterns, perforations, shapes of deformers, changing angles and rotation of the robot target path, amount of deformation, amount of heat applied, speed and timing of the procedure.

3.1 Joint Between Frame and Robot

To allow a maximum degree of freedom for the movement of the panel, the frame is attached to the robot's 6th axis by a T-joint. Therefore, during one procedure, complementing the freedom of the robotic arm movement, the frame can rotate horizontally 360° (Fig. 4). Deformations can be applied to all sides of the surface within one process.

3.2 Frame and Panel

In this case, the frame is the custom end-effector. It is laser cut from wood and attached to the robot via the T-joint. Frames of different shapes (Fig. 5, left) can easily be exchanged. The frame consists of an upper and a lower part, between which the overlapping panels are placed and fixed.

In the frame used in this study, panels with a maximum size of 30 by 30 cm can be placed. The same frame can also be used for panels with other fitting geometry, i.e. rectangular or triangular in a square frame (Fig. 5, right). The clamped edges stay straight and the free edges can be deformed.

3.3 Laser Cutting

Before thermoforming, slots or openings are laser-cut to create apertures. By scoring lines or patterns on the surface, transparencies and surface qualities can be altered. Individual patterns are designed for each panel according to the qualities that shall be achieved after completion of the fabrication process.

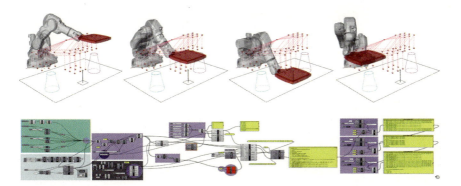

Fig. 3 Screenshots of Rhino/Grasshopper/HAL: parametric setup of the robotic process, path simulation and code generation

Fig. 4 Joint Frame/Robot: rotation of frame via the 6th axis of the robot

Fig. 5 *Left* Selection of possible frame shapes; *Right* Relation frame/possible panel shapes

3.4 Deformer

The heated panels are robotically shaped by pushing them against a "deformer". Variable geometries and sizes of deformers can be used. In this prototype-series, it consists of a rounded wood sphere with 5 cm diameter.

With the use of parametric programming of the robot procedures, each panel can be deformed at any number and positions of target points. In this set-up a series of nine sets of target points is used, each moving from the heat gun, to the deformer, and to the cold air fan. After being heated, each panel is deformed at nine points by the impression of the deformer.

3.5 Hot Air, Cold Air and Wait Time

Before deformation, the panel has to be heated-up to a temperature that is high enough to allow the deformation, but not too high, so that the panel deforms only by gravity. The first target point of each deformation set is above the heat gun. In this fabrication series, the panels are placed 10–15 cm from the top of the heat gun, which is turned on to maximum power. The temperature is defined by the "wait time", the amount of time the panel stops above the hot air. In this set-up, it is between 20 and 60 s.

At the peak of each deformation, when pushed against the deformer to the maximum depth, the material is cooled down to a temperature that allows the applied deformation to retain its shape. This is achieved by implementing another wait time, when the movement is suspended at the position for 5–15 s. Subsequently, the panel is moved above the cold air fan for another 30–60 s.

3.6 Robotic Process

The robotic process is a very precise one, but still there are many empirical factors that affect the outcome of this deformation-process.

For the different panels, changes in the parametric set-up of the robot path are made according to the location of the target points, based on the position of the pre-cut openings or slots and the desired depth of deformation. The "wait times" at each target are altered, defining the amount of heating and cooling. Also, the actual moving speed of the panel when pushed against the deformer needs to be adjusted. The larger the deformation, the slower the panel should move.

After the RAPID-code for the machine is generated in HAL, the robot movement is simulated and double-checked in ABB RobotStudio.

4 Prototypes

In this chapter, a selection of the produced prototypes, made from acrylic glass in the size of 30 by 30 cm and 3 mm thickness, is shown. The different modules are formed making small adjustments to the same procedure to produce varying shapes, apertures and textures.

4.1 Prototype A: Continuous Linear Cuts

For this prototype, a sheet of transparent acrylic glass is cut to size and laser slotted along parallel lines extending from one end to the other end of the panel. Interesting findings in the production are that, due to the air-permeability of the sheet, the heating time before deformation can be very short. In the forming process,

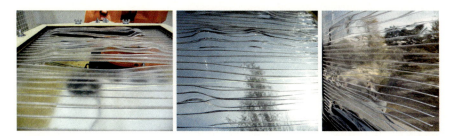

Fig. 6 Prototype A: during production and applied as facade panel

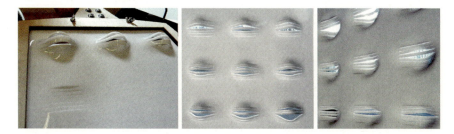

Fig. 7 Prototype B: during production and applied as facade panel

each target point along the same cutting line influences the deformation of the next point, creating apertures in varying size. Applying this prototype as facade panel, different shades of transparency and reflection become visible, depending on light conditions, view-angle and apertures (Fig. 6).

4.2 Prototype B: Segmented Linear Cuts

In this case, segmented parallel slots are cut into opaque acrylic glass. The heating time needs to be quite long due to the ductility of the surface that has only minimal cuts. Before thermoforming, the cuts are hardly visible and the surface is closed. Through the deformation, the cuts stretch and open the previously opaque surface. Apertures are created along the slots, allowing light and air-permeability, contrasting between opaque and transparent qualities of the surface (Fig. 7).

4.3 Prototype C: Diamond-Shaped Cut-Outs and Diamond-Shaped Grid

For this prototype, made from a transparent sheet, diamond-shaped openings are cut, and a matching grid pattern is scored to change the transparency and surface quality of the panel. During the deformation process, the scored grid breaks up and

Fig. 8 Prototype C: during production and applied as facade panel

Fig. 9 Prototype D: during production and applied as facade panel

the perforations change their shape to almost elliptical apertures. The appearance of the module is fabric-like. In the application of this prototype as facade panel it can be observed, that it creates a filtering and texturing effect (Fig. 8).

4.4 Prototype D: Triangular Cut-Outs in Varying Sizes

Here, an almost opaque panel is perforated by subtracting triangles in varying size, influenced by the distance to the points of deformation. In this process, two different sizes of spherical deformers in diameter 3.5 and 5 cm are used, which are exchanged at every second point. In this prototype it becomes visible that, responding to the deformation, the openings get stretched differently, or, at the minimal size, even melt together (Fig. 9).

5 Conclusion

As shown with this series of prototypes, intriguing and individual shapes can be generated by this novel fabrication technique, combining laser cutting with robotic deformation.

The main advantage to other forming techniques like vacuum forming, where the shape needs to be defined with rigid tooling, is that the technique developed in this research is a dynamic shaping process - a flexible process from design to production, where to outcome is not pre-defined. Together with other main factors -which are the material properties, the shape of the deformer and the cuts in the panels - the movement of the robotic arm creates the final form. The robotic procedure is part of the form giving process: it is a "dynamic mold".

The next step in the line of this research will be the optimization of the prototypes for specific site or program conditions. Possible applications in architecture are building skins, additive or multi-layered facades or systems of primary skin structures. The technique will be explored further using other materials suitable for architectural applications, like bio-plastics or smart materials.

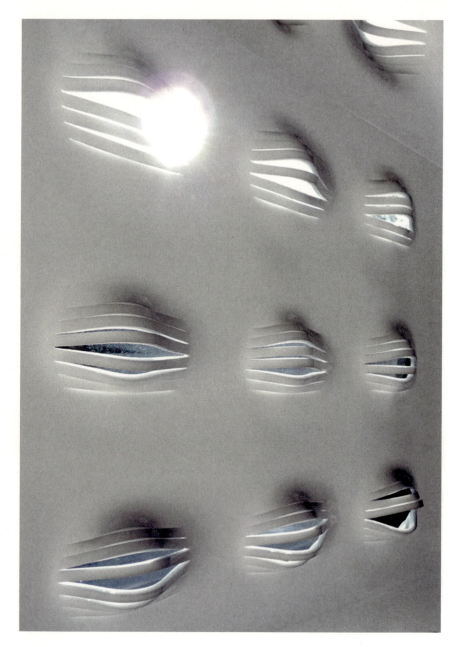

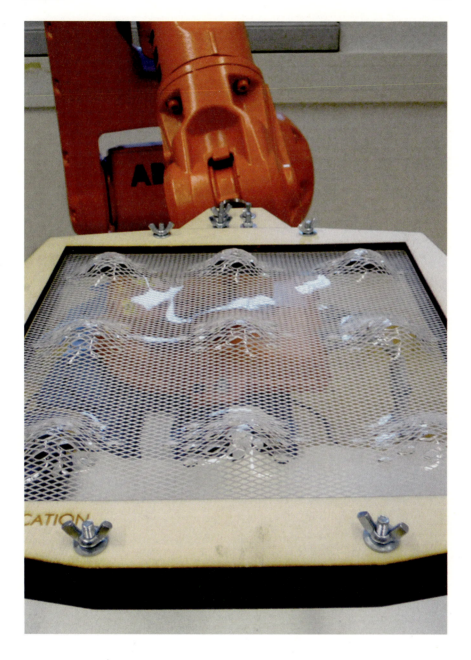

References

Brell-Cokcan S, Braumann J (eds) (2012) Rob|Arch robotic fabrication in architecture art and design. Springer, Vienna

Hensel M (2013) Performance-oriented architecture: rethinking architectural design and the built environment. Wiley, UK

Riether G, Jolly K (2011) Flexible systems: flexible design, material and fabrication: the AIA pavilion as a case study. In: Proceedings of the eCAADe conference, Ljubljana, Slovenia, pp 628–634

Experiments in Additive Clay Depositions

Woven Clay

Jared Friedman, Heamin Kim and Olga Mesa

Abstract The Standard definitions and models of additive manufacturing, such as those outlined by ASTM (2012), often assume a layer-by-layer deposition of a material onto a flat surface. The research presented looks to explore what alternative formal opportunities that may arise when challenging these assumptions concerning additive manufacturing. Beginning with the robotic clay coil extrusion process, the research uses a bottom–up approach that asks what types of forms and scales can be generated with this process. Prototypes incorporate techniques such as weaving as a means of fabricating panels that celebrate the inherent nature of the coil itself. The woven patterns are designed to incorporate both assembly logics and performance qualities such as light permeability across a façade component. Through a series of full-scale prototypes focusing on a specific building application, this research works to resolve the seemingly conflicting nature between the inherent mutability of the clay material and the high level of control granted by robotic fabrication processes.

Keywords Additive manufacturing · Ceramics · Clay deposition · Weaving

J. Friedman (✉) · H. Kim · O. Mesa
Harvard Graduate School of Design, Cambridge, MA, USA
e-mail: friedman@gsd.harvard.edu

H. Kim
e-mail: hkim@gsd.harvard.edu

O. Mesa
e-mail: omesa@gsd.harvard.edu

W. McGee and M. Ponce de Leon (eds.), *Robotic Fabrication in Architecture, Art and Design 2014*, DOI: 10.1007/978-3-319-04663-1_18,
© Springer International Publishing Switzerland 2014

1 Introduction

The research presented in this chapter looks towards utilizing the additive clay coil extrusion process as a means of fabricating panels that can be applied as architectural screens or façade components. Early explorations began by looking at historic applications of woven screens in architectural settings, as well as the application of clay coils that have typically been applied to the making of bowls and pots. The research proposes woven robotic clay deposition as a means of fabricating architectural panels whose forms are driven by the modulation and filtration of light.

ASTM (2012) defines "additive manufacturing" as the "process of joining materials to make objects from 3D model data, usually layer upon layer." By challenging the "layer upon layer" printing techniques used in typical 3D printing processes, new geometries and surface patterns may be achieved with alternative deposition strategies. Using the traditional clay-coil extrusion as a formal language, woven deposition techniques may be utilized as a means of achieving various lattice patterning effects along the surface of the printing bed. These patterning effects leverage inherent properties of the clay, while also taking advantage of the control, size, and speed granted by the use of a 6-axis industrial robot.

The first prototypes developed all utilized manual methods of clay extrusion, which then informed the digital and physical tooling that would be necessary to transfer the process to robotic fabrication methods. Robotic clay deposition allowed for quicker and more consistent extrusions than were achievable with manual methods. Additionally, use of the digital tools Grasshopper®, developed by Scott Davidson, and HAL Robot Programming and Control, developed by Thibault Schwartz enabled a highly customizable and automated workflow that self-adjusts based upon the input geometry. The research outlined in this chapter places the project within a larger context, while describing the design and fabrication methods that have been applied throughout the process.

2 Precedent Analysis

2.1 Traditional Applications of Ceramics and Weaving

Traditionally ceramics have been used in architectural applications as a means of formal expression as well as a filter of light, sound, and smell. This type of application is exemplified in projects such as the Alhambra palaces in Granada, where ceramic screens are applied as a dynamic filter that mediate between two separate spaces (Lazcano 2008). At a micro-level, clay techniques involving the assembly of coils is one that demands lots of time and precision, yet is still one of the oldest and most universal ceramic techniques (Peterson and Peterson 2003). The time and precision needed for this process lends itself well towards robotic manufacturing applications.

The use of weaving as a technique to create walls or screens has been pervasive throughout architectural history—from the most primitive wickerwork fences made of sticks to the most intricately designed tapestries woven in silk. Gottfried Semper ([1851] 1989) describes the important role these techniques play in their application as the earliest expressions of the wall as an architectural element. Semper ([1862] 1989) also discusses the ways in which the styles of textiles have been influenced by various factors including the influence of machine fabrication. This relates directly to how the organizational logic inherent in textile patterning lends itself to the translation into the digital realm—a theme emphasized more recently by Lars Spuybroek (2011) in his writings on textile tectonics.

2.2 Current Models of Robotic Clay Deposition

Much of the recent research in robotic ceramic deposition techniques has looked to capitalize on the customization and precision granted by the process. Research conducted by the Design Robotics Group at the Harvard Graduate School of Design looked at ceramic deposition as a method to fabricate custom louvers for ceramic shading systems (Bechthold et al. 2011). While the research looked at deposition over a non-flat printing base, the process is limiting in that it is used to print a continuous surface that attempts to mask the scalloping that occurs as a result of the deposition technique. Similarly, research conducted by Berokh Koshnevis (2001) on 'Contour Crafting' proposes the use of a trowel to smooth printed walls in order to obtain smooth planar and freeform surfaces. Rather than attempting to mask or smooth the striations that result from the process, our research looks to embrace the variability of the material and deposition process (Fig. 1).

2.3 3D Printing with Ceramics

Another more recent application of ceramic deposition has stemmed from customized RepRap (reprap.org/wiki/RepRap) machines that have developed pneumatic deposition techniques for printing smaller scale components in a highly controlled manner. These precedents include projects such as Building Bytes by Brian Peters (buildingbytes.info/) and ceramic printing studies conducted by the Belgian design studio Unfold (unfoldfab.blogspot.com). Since ceramic printing techniques cannot utilize the same disposable support materials seen in other printing techniques, many of these precedents are formally limited to modules with flat tops and bottoms. While this aspect of the unit offers the advantage of having the modules to be self-supporting, there is little evidence showing that these blocks would provide significant load bearing capacities at large scales.

A primary limitation in 3D printing projects proposing standard brick-sized components is that the due to the machine time necessary to produce enough

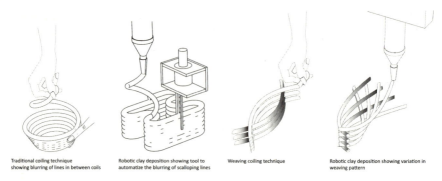

Traditional coiling technique showing blurring of lines in between coils

Robotic clay deposition showing tool to automatize the blurring of scalloping lines

Weaving coiling technique

Robotic clay deposition showing variation in weaving pattern

Fig. 1 Typological diagramming of precedents and proposed method

Table 1 Comparing machine times for building bytes module versus printed ceramics panel

Unit type	Printing time (per unit) (min)	Quantity of units (per 100 sq. ft.)	Machine time (per 100 sq. ft.) (h)
Building bytes block	15	450	112.5
Prototyped ceramic panel	8	50 (after 12 % shrinkage)	6.67

components to build up a surface, it is neither economical nor efficient at a large scale. Table 1 compares machine times for the Building Bytes module at 15 min a block (www.dezeen.com/2012/10/31/building-bytes-3D-printed-bricks-brian-peters/) for a standard brick size versus the panels prototyped during this research, which average 8 min for an 18″ × 18″ panel. Machine times for the molds used for the panels are not taken into account here due to the fact that at a production scale, molds can be reused, and machine time would become increasingly insignificant over time. Ways to increase the practical application of ceramic printing processes include producing larger components at a more rapid pace.

3 Physical Tooling and Fabrication

3.1 Material Selection

Numerous clay bodies were tested in order to find one with ideal properties for the research. The clay needs to have enough plasticity to easily be pushed out of the extruder at a constant speed, while still providing the stiffness necessary to retain the coil shape. Shrinkage rates are also critical to consider. Clays with lower shrinkage rates are less prone to cracking while drying, and are more predictable since they contain less moisture at the time of deposition. While developing our own clay mixtures allowed for a high degree of control over the plasticity and shrinkage, it was often a time consuming process that resulted in batches of clay with lots of air bubbles. The authors' studies showed that using clay directly from the pug mill was much less prone to air bubbles, which reduced coil breakage, and allowed for the use of off-the-shelf clay mixtures. After testing a number of clay bodies directly from the mill, a Cone 6 Porcelain clay body was selected for its consistency and for having the necessary plasticity due to the fine particle size of porcelain clays. The Cone 6 Porcelain provided a 12 % shrinkage rate, which was one of the lowest shrinkage rates among the clays that were tested.

3.2 Robotic Tooling

Some of the key variables identified in relation to the robot tooling were the diameter of the extruded coils and the speed at which the clay was driven from the extruder. The extruder utilized, developed by the Design Robotics group at Harvard GSD, operates via mechanical means with a gear motor that drives a lead screw into a plunger that pushes the clay through a custom nozzle (King et al. 2011). A removable canister holds the clay, which is loaded before each run. A nozzle was made that would provide a coil diameter of 3/8″, which was selected for the resolution and speed that it allowed for when printing a panel that is roughly 18″ × 18″. This size panel was selected for the prototypes based upon parameters including quantity of clay the canister can hold, dimension of the firing kiln used, and ease of assembly by a group of students.

3.3 Robot Movements and Toolpath Generation

A number of variables related to the robot's movement were evaluated in order to achieve the most accurate clay deposition relative to the input curves the robot was able to follow. Toolpaths generated using Grasshopper® components used a B-spline curve as the centerline of the toolpath. The curve had to be offset from the printing surface at a specified height in order to provide the most accurate and consistent results. Since the behavior of the clay deposition varies depending on the degree of curvature, it is necessary to calculate the curvature at each point passed to the robot, and link a specific speed to that point so that the robot will move more rapidly over areas of high curvature and more slowly over areas of low curvature (Fig. 2).

Once a desired pattern has been determined based on aesthetics and porosity levels, a specific logic is applied to the printing paths in order to maximize overlaps and interweaving—thereby decreasing the fragility of the panels (Fig. 3). A single layer of coil is deposited along one continuous path that overshoots the edges of the panel. This is done to reduce time spent stopping and cutting coils, while also allowing the clay coils to catch the edges of the mold so that they don't slide or shift. All excess clay cut off of the edges at the end is able to be reused.

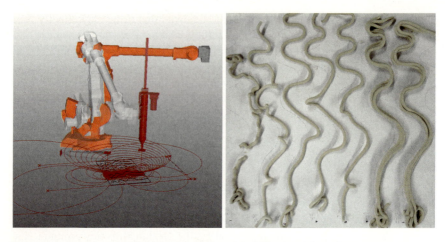

Fig. 2 Simulations run in HAL and the testing of variables involving robot movements

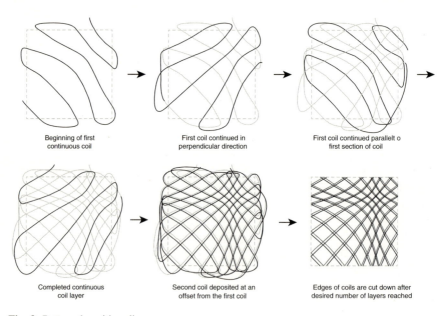

Beginning of first
continuous coil

First coil continued in
perpendicular direction

First coil continued parallelt o
first section of coil

Completed continuous
coil layer

Second coil deposited at an
offset from the first coil

Edges of coils are cut down after
desired number of layers reached

Fig. 3 Pattern deposition diagram

Throughout the prototyping process, various milled foam molds were used as a means of testing various surface types. A key observation made through the prototyping process was that a large degree of variation is possible within a single mold (Fig. 4). Based upon the deposition pattern and the density of the coils, one can achieve a wide range of opacities and visual effects (Figs. 5 and 6).

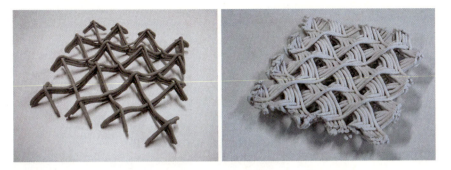

Fig. 4 Two handmade prototypes done on the same base mold

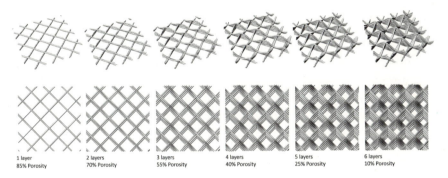

1 layer
85% Porosity

2 layers
70% Porosity

3 layers
55% Porosity

4 layers
40% Porosity

5 layers
25% Porosity

6 layers
10% Porosity

Fig. 5 Varying opacities over a single mold

Fig. 6 Robotically printed panels prior to firing

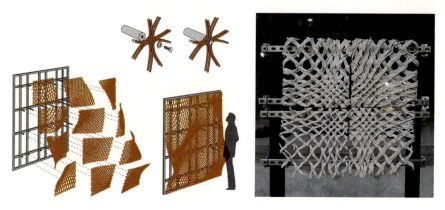

Fig. 7 Diagram and photograph of assembly system on a building façade

4 Digital Workflow

The first step of the digital process is to generate surfaces that act as the bases for the coil deposition. The prototypes produced utilized Grasshopper® to develop-bumped surfaces from sine curves to emphasize the variation in depth. Once the patterns are generated along the surface, they are turned into toolpaths using Grasshopper® and HAL components. To generate the RAPID code for each pattern, one simply reselects the curves that are intended to become the toolpaths. This process allows for a quick design-to-fabrication workflow in order to decrease the time devoted to each panel. In a larger application, a workflow can be developed that would allow the designer to select features such as the desired opacity across a surface at the front-end of the workflow.

5 Industrial Applications and Future Development

The argument for this process to be applied at a larger scale application is driven by the speed of fabrication in relation to the size of the panel that can be produced. While the prototypes produced for this research were limited based upon the quantity of clay the tool can hold and the dimensions of the firing kilns used, in an industrial setting these would not be the primary limitations. In industrial settings limitations are more likely to come from span limitation of the ceramic material, which would also depend on the assembly method. Possible applications include use of the panels as shading mechanisms that take into consideration various environmental factors such as solar shading along a façade.

The research performed throughout this project has displayed the potential to expand beyond commonly accepted practices within additive manufacturing. Through the use of weaving deposition techniques onto a non-flat surface it is possible to begin to challenge the formal assumptions typical to the 3D printing of ceramics. One area requiring further development is the assembly logics, and how the patterns and edges of the panels are impacted by the assembly (Fig. 7). Further exploration is needed to successfully resolve these conditions. While the primary intention is to investigate new opportunities granted by the process, it is also critical to consider the reliability of the process within other manufacturing settings. The level of customization, scale, and speed at which the panels were produced is a testament to the potential of the process for large scale installations.

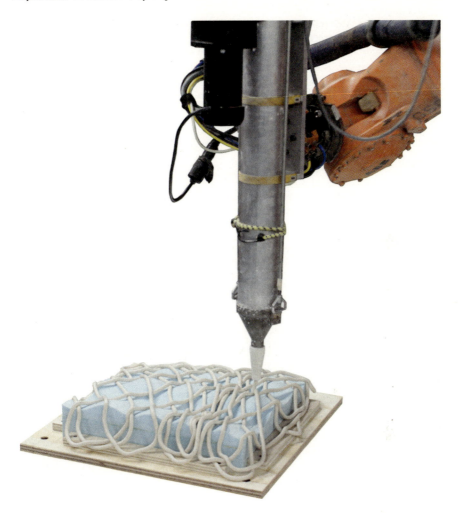

Acknowledgments This research was conducted under the guidance of instructors Nathan King and Rachel Vroman during the course: Material Systems: Digital Design, Fabrication, and Research Methods at the Harvard University Graduate School of Design; Cambridge MA; Fall 2013.

Research supported by ASCER Tile of Spain; Harvard University Graduate School of Design, Design Robotics Group; the Office for the Arts at Harvard, Ceramics Program; and the Harvard Graduate School of Design, Fabrication Laboratory.

References

ASTM (2012) F2792-12a Standard terminology for additive manufacturing technologies. ASTM International, West Conshohocken, PA

Bechthold M, King N, Kane A, Niemasz J, Reinhart C (2011) Integrated environmental design and robotic fabrication workflow for ceramic shading systems. In: Proceedings of the 28th international symposium on automation and robotics in construction. Seoul, Korea

Lazcano J (2008) Ceramics in the Alhambra: petrified water. Essays Archit Ceram 4:67–77

King N, Bechthold M, Kane A (2011) Customizing ceramics: automation strategies for robotic fabrication. Pending Publication, Digital Futures Tongji University Press

Koshnevis B (2001) Experimental investigation of contour crafting using ceramic materials. Rapid Prototyp J 7(1):32–41

Semper G (1851) 1989 The fours elements of architecture'. In: The four elements of architecture and other writings. Cambridge University Press, Cambridge

Semper G (1862) 1989 Style in the technical and tectonic arts or practical aesthetics'. In: The four elements of architecture and other writings. Cambridge University Press, Cambridge

Spuybroek L (2011) Textile tectonics. NAi Publishers, Rotterdam

Peterson S, Peterson J (2003) Craft and the art of clay: a complete potter's handbook, 4th edn. Lawrence King Publishing Ltd., London, pp 33–34

Part III
Workshops

Core-Less Filament Winding

Robotically Fabricated Fiber Composite Building Components

Marshall Prado, Moritz Dörstelmann, Tobias Schwinn, Achim Menges and Jan Knippers

Abstract The research presented in this chapter describes novel strategies towards robotic fabrication of geometrically complex fiber reinforced building elements. The research focuses on "core-less" filament winding processes which reduce the need for formwork allowing for the fabrication of individual one off components with differentiated fiber layout. The first part of the chapter introduces the need for advanced fabrication strategies in order to use the full potential of fiber composites anisotropic material behavior and the need for complex geometries in performative lightweight structures. The second part contextualizes the presented work by linking it to relevant contemporary and historical precedent. The main part of the chapter discusses methods developed for the "core-less" filament winding processes, followed by conclusions and outlook towards future potentials.

Keywords Robotic fabrication · Fiber composites · Core-less filament winding · Lightweight

1 Introduction

Fiber composites open up new possibilities for material experimentation in architectural design and construction through an inherent capacity to form complex geometries and programmable anisotropic material behavior. The explorations,

M. Prado (✉) · M. Dörstelmann · T. Schwinn · A. Menges
Institute for Computational Design (ICD), University of Stuttgart, Stuttgart, Germany
e-mail: mail@icd.uni-stuttgart.de

J. Knippers
Institute for Building Structures and Structural Design (ITKE), University of Stuttgart, Stuttgart, Germany
e-mail: info@itke.uni-stuttgart.de

W. McGee and M. Ponce de Leon (eds.), *Robotic Fabrication in Architecture, Art and Design 2014*, DOI: 10.1007/978-3-319-04663-1_19,
© Springer International Publishing Switzerland 2014

conducted as part of the research into robotic fabrication, are a result of design strategies at the scale of material organization and allow for a higher degree of material efficiency and morphologic articulation in lightweight constructions. Such a system requires the fabrication of complex geometries with precisely controlled fiber orientation—a difficult process to achieve on an architectural scale using traditional modes of fiber-reinforced polymer (FRP) fabrication. Therefore novel processes for automated fabrication are explored to increase both control of fiber placement and geometric versatility compared to existing modes of fabrication. A "core-less" winding approach was developed which creates component morphology through a designed sequence of fiber–fiber interaction (FFI) rather than the need for prefabricated winding mandrels or costly formwork. The goal of this research can be summarized as follows: to integrate the various requirements of design, analysis, and fabrication, i.e. coreless robotic filament winding, into one coherent design framework for the design of lightweight, materially and structurally efficient fiber-reinforced building components (Fig. 1).

2 Context

Fiber-reinforced polymers possess a number of characteristics such as a high load-bearing capacity combined with minimal self-weight, which have made them the preferred choice for structural applications in a variety of engineering fields ranging from automotive to ship-building to aerospace. Existing fabrication techniques can be summarized in a handful of distinct processes such as pultrusion or lay-up of fibers. The fibers (usually in the form of mats, pre-pregs, or rovings) are laid onto prefabricated formwork using various degrees of automation ranging from hand layup to robotic fiber placement. Not only can fabricating structural elements using FRP be surprisingly low-tech, requiring no industrial-scale equipment, but also CNC-controlled machinery for producing geometrically complex formwork has become ubiquitous in recent years, and so the race for ever higher performing applications of FRPs has led to a culture of experimentation, spin-offs and startups (Anderson 2012). This exciting development begs the question why developments of structural applications in architecture seem to have stalled after an initial phase of experimentation in the 1960s and 70s (Voigt 2007).

Performative applications of FRP, i.e. applications that take advantage of the material's inherent structural capacity, demand that the geometry of the structural element as well as the directionality of the fibers in its laminate be optimized for the specific structural requirements. The resulting differentiation of geometry has to be matched by an appropriate manufacturing process that allows for an economic fabrication: batch to mass production utilizing reusable molds in the case of automotive and aerospace, or adaptable formwork in the case of one-off high-performing sail manufacturing (Gustafson 2010). This hints at a fundamental challenge faced by a structural application of FRPs in an architectural context: an economy of scale where the initial cost of using custom molds is offset by the

Fig. 1 a 12 axis robotic setup. **b** Image of fiber composite building component

serial production of geometrically complex, albeit identical elements; or, as in the example of manufacturing high-performing sails, initial investment in complex actuated formwork, which allows variation within the defined limits of the application, is enabled by an exclusive niche market. Neither of these strategies seems applicable to the building industry. The geometric differentiation necessary to take advantage of the material's capacity, on the component level as well as on the level of the assembly, combined with the scale and the sheer number of elements usually makes one-off solutions impractical.

Nevertheless, a renewed sense of experimentation seems to have leaped over to architecture from these other fields as a series of recent innovative architectural prototypes using FRP indicate. These examples intrinsically also provide an indication of the inherent economics of reusable molds versus geometric differentiation. Examples include (1) the pavilion at the South Pond of Lincoln Park Zoo in Chicago by Studio Gang (2009), (2) the Cocoon_FS pavilion by Pohl Architekten (2011), (3) the Chanel Mobile Art Pavilion by ZahaHadid Architects (2010), and (4) the ICD/ITKE Research Pavilion 2012 at University of Stuttgart. While the South Pond pavilion is clad with identical, semi-transparent fiberglass elements for shading and shelter (Gang 2010), the Cocoon_FS pavilion is composed of a series of geometrically differentiated elements. Still, the project's stated goal was the reduction of the number of unique elements and, consequently, the amount of differentiation in response to fabrication considerations. Therefore the amount of unique molds is kept to a minimum (15 types within an overall number of 220 panels) (Fischer 2012). The facade of the Chanel Mobile Art Pavilion, however, is composed of 400 geometrically unique glass fiber reinforced elements (Stage One 2009). While all three examples are formally and functionally very distinct, they all fundamentally share the same traditional fabrication process of fiber layup in prefabricated formwork. In this process, the fabrication expense, which includes the material and machining of the mold, is proportional to the number of unique elements, which, consequently and unsurprisingly, has the Chanel Mobile Art pavilion lead the field in terms of construction budget. The ICD/ITKE Research Pavilion

2012 project, on the other hand, identifies the need for the reduction or even elimination of formwork in the production of performative FRPs on the architectural scale. This project is characterized by its unique fabrication process of robotically winding glass and carbon rovings over a prefabricated lightweight steel frame (Reichert et al. 2014).

Advances in fabrication technology, and specifically the increasingly widespread use of industrial robot arms, play an important role in the development and implementation of such a novel fabrication approach. The degrees of freedom of a 6-axis robot arm combined with its relative precision provide opportunities with regards to adapting building elements to the specific localized requirements they must meet in an assembly: e.g. component geometry and fiber orientation in response to loading and structural requirements. In order to activate this potential, innovative approaches to robotic control are required, including toolpath generation, simulation of robot kinematics, and code generation for 6 + axis robot systems.

Over the course of approximately the last 5 years, field-specific developments regarding robot control have taken place in architectural design and digital fabrication research. These include, for example, the development of production immanent design tools that increase feedback between fabrication constraints and high-level design decisions by eliminating the traditional production sequence from computer-aided design (CAD) to computer-aided manufacturing (CAM) to fabrication (Brell-Çokcan and Braumann 2010); the development of application-specific CAM strategies, e.g. for finger-jointing of plywood plates (Schwinn et al. 2012); but also innovations in man–machine interaction and the integration of sensors in the production process. These developments aim not only at making robot fabrication more intuitive, and, ultimately, more powerful; they also drive innovation and expand on existing processes in specific fields, such as timber construction or FRP production, through robotic fabrication.

3 Methods

In the ICD/ITKE Research Pavilion 2012, a single robotic arm with an external rotational axis was used to create a light-weight monocoque structure. The size and morphology of the pavilion was a function of the reach of the robot arm and the particular 7-axis setup using an external vertical axis as turntable. To overcome the limitations of such a particular setup, a structure that involves multiple smaller scale, yet highly differentiated building components was proposed, which in turn requires a significantly different fabrication setup.

3.1 Physical Tests

Physical modeling, which was utilized in several ways during the course of the project, was an empirical means of testing and adjusting digital design and fabrication strategies. First, physical modeling was used to test spatial design strategies. These included geometric variations of the components such as shape, size and orientation. Component geometry can be minimally characterized by two polygons that are defined by parameters determined from the empirical test (Fig. 2). Second, physical models were used to test the winding syntax, i.e. the systematic sequence of fiber placement. This is represented graphically by a continuous polyline or denoted textually by a code listing the points in the order they are wrapped (i.e. A16, B32, A48, B60, A75…). The winding syntax should result in good fiber–fiber interaction (FFI). The composite structure is dependent on fibers pressing against previously wound fibers, insuring that the system remains tensioned. This fiber behavior relies on the coordination of the frame geometry, winding locations and sequence. Fibers connecting control points would initially define a straight line but as subsequent fibers are wrapped, a hyperbolic surface emerges defined by fiber tension and the order in which they are wrapped (Fig. 3). Fibers can also be differentiated through anisotropic fiber placement. Winding fibers along, or intersecting at, major stress directions increases material efficiency and structural performance. Lastly, hand winding physical models is analogous to a robotic fabrication process and could be used to test potential problems of automation, but it also illustrated the need for an integrated design tool which considers material behavior, structural performance and fabrication constraints for the design and fabrication of coreless fiber composite building components.

3.2 Fabrication Setup

In order to alleviate the need for the core or mandrel in the winding process it was necessary to develop a fabrication system in which fibers could be wrapped around minimized reconfigurable scaffolding. This complex, non-planar frame geometry needed to be accurately fixed in space to insure the precision of the finished wrapped building component. The cumulative tensioning of each fiber produces large forces distributed across the surface, acting on the scaffolding. It became necessary to develop a robotic fabrication process in order to accurately define the frame geometry in space, wrap a large number of points quickly and efficiently, and resist the internal forces applied to the system. Therefore a robotic controlled assembly process, consisting of multiple adjustable parts, was implemented that allows the accurate definition the non-planar polygon for each component while providing the stability required to withstand the tension forces during the winding. An initial setup with a single robotic arm limited geometric freedom or required a

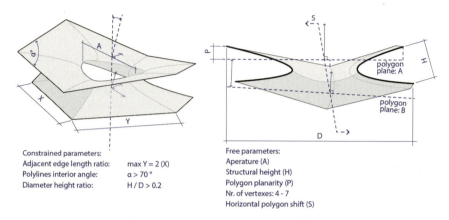

Constrained parameters:
Adjacent edge length ratio: max Y = 2 (X)
Polylines interior angle: α > 70 °
Diameter height ratio: H / D > 0.2

Free parameters:
Aperature (A)
Structural height (H)
Polygon planarity (P)
Nr. of vertexes: 4 - 7
Horizontal polygon shift (S)

Fig. 2 Diagram of the component constraints derived through empirical tests

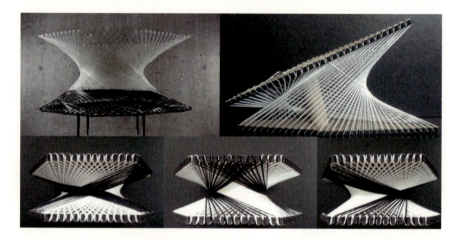

Fig. 3 Hand wound models of the winding syntax, FFI and differentiated fiber layout

more involved reconfigurable framework. The complexity in the framework could be reduced by shifting to a 12 axis robotic setup.

A 7 m × 7 m × 7 m production space was equipped with a robotic cell containing two synchronized Kuka KR 210 R3100 robots for the winding of the elements, an assembly stand for the robot assisted assembly of the effectors and a resin impregnation bath. Each robot's effector consisted of a 2 m diameter octagonal steel truss to resist the tensile forces of the fibers and provided a rigid, yet light-weight support for the reconfigurable frame stands to be attached (Fig. 4). The frame stands were designed to be robotically assembled for various non-planar components using the effector assembly table which was positioned at a central location in the workspace between the two robots. By robotically surveying points on the resin bath, assembly table, and steel trusses, the base planes and tool planes on each robot could be determined. Referencing of the physical fabrication environment by the robots is one of many crucial steps to help assuage discrepancies between the digital model used for simulation and code generation, and the physical fabrication setup.

The precise location of the wrapping tool in space is achieved by a robot-controlled positioning process. The robot positions each steel frame effector in space in relation to the effector assembly table location. The frame stands, which have multiple degrees of freedom for fabrication yet provide the rigidity and precision for fiber winding, are then manually fastened to the effector. This combination of robotic precision and fast manual fastening works as follows: a 5-axis milled plywood plate, which has two drilled reference holes, is affixed to the assembly table which has two corresponding surveyed pin locations to define the base plane; the robot is rotated into place so that it is in the correct corresponding location and orientation; the adjustable metal legs of the frame stands are then secured to the effector defining one vertex of the component polygon (Fig. 5). These steps are repeated for each vertex in the polygon and on each effector. Adjustable bars, containing control points spaced every 4 cm, are attached to the frame stands and define each segment of the polygon. These bars are held in place by the wood plates which are custom fabricated for each component and define the orientation plane at each vertex as well as the rotation of each bar (Fig. 5). The attachment locations and rotation for each bar are encoded in the wood plate during the milling process to ensure alignment from one component to the next (Fig. 6). This also serves as an error check in the assembly process since the bars cannot be attached to misaligned wood connectors. The robotically controlled effector assembly process assures that despite any inaccuracies in the construction of the steel truss or metal legs, the wooden connectors and by virtue the bars and control points will be properly oriented for winding with respect to the robot.

3.3 Robot Synchronization

With two fully assembled effectors, the robots are synchronized, effectively creating a 12-axis kinematic system, to insure geometric accuracy of the effectors and coupling for winding. One robot is designated as the master and one, as the slave. In order for the two robots to be synced the slave follows a base plane that is kinematically linked to the master. The offset and orientation of the slave is programmed from the geometric model by automatically changing the stored variables of the linked kinematic base when the slave coupling program is activated. With the stored variables changed the slave moves into the proper orientation before awaiting further instructions from the master. After the master coupling program is activated and synchronization is established the robotic winding path for both robots can be processed through the code running on the master.

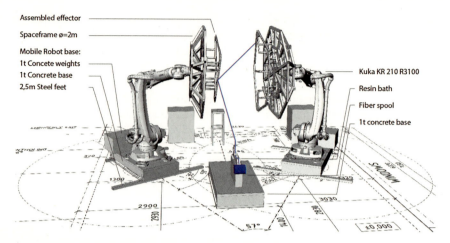

Fig. 4 Diagram of the roboticsetup

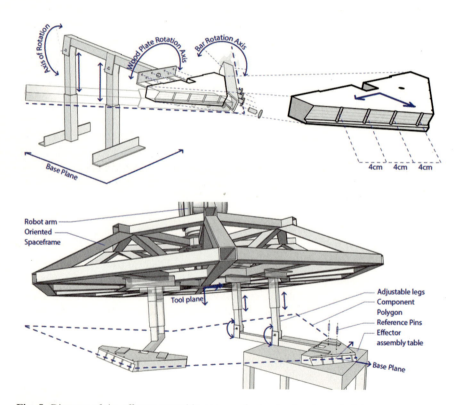

Fig. 5 Diagram of the effector assembly process, frame stand and wood plates

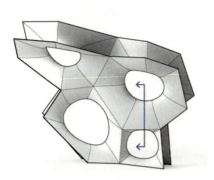
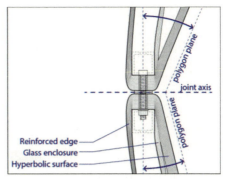

Fig. 6 Component connection detail

3.4 Winding

There are three steps used to develop robotic winding codes. These include the development of a winding syntax, digital simulation of the robotic movements and code generation of a robotic winding path. For the purposes of this project all three interdependent processes were developed as custom scripted routines.

Each component produced was wrapped with 6–7 winding patterns. The first is an enclosure layer representing the outside finished surface of the structure, which is wrapped only on the master effector. This pattern is wrapped in glass fiber to approximate a spatial enclosure and is neither intended to provide protection from the elements nor structural reinforcement. The second layer is the substructure layer of glass fiber which connects the master and the slave effectors. This layer, which is similar on all components, is used to define the overall geometry of the component on which the carbon fiber is wrapped. In this way it serves as a "lost fibrous mandrel" for the subsequent layers of wrapped fiber. The structurally differentiated carbon fiber is then wrapped between master and slave reinforcing specific areas of high stress. Next, a generic layer of carbon fiber is wrapped on each frame to compress all layers into a pre-tensioned composite. A final layer of carbon fiber is then added to the outside of each frame to reinforce the edge connections (Fig. 7).

It is important to note that the winding syntax polyline, the wound fiber roving that it sequentially represents and the robotic winding path used for fabrication, are all very different representations of how fibers are wrapping each component (Fig. 8). The wound rovings which follow the winding syntax create a curved line connecting fibers from control point to control point. The robotic winding path, which is derived from the winding syntax, controls the robotic arm to avoid collisions and singularities in robotic movement whilst delivering fiber to the defined locations. Looping motions are inserted at every control point to guide the fiber source around the hooking area. Fine adjustments of the robotic winding paths, base plane orientation and centers of rotation for each robot are necessary

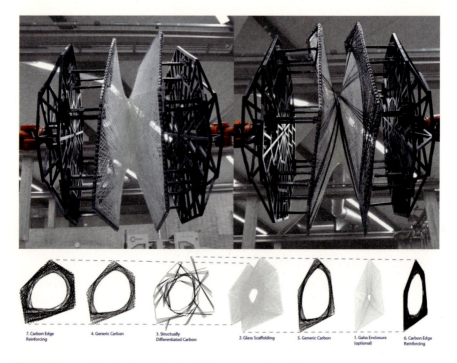

Fig. 7 Diagram showing the order of winding patterns for each component

when extreme component size or non-planarity creates robotic movements closer to the minimum/maximum reach of the robot or raises the possibility for the effector frame to collide the robot or the resin bath. Evolutionary algorithms were used to optimize these parameters to avoid potential winding problems. The simulation used a custom scripted inverse kinematic solver for 12 axis robotic simulation. A winding base plane on the resin bath was defined similar to the base plane on the effector assembly table. Planes defining the robot location and orientation, at each winding point, control where the robot moves to in space. Each line of the winding syntax polyline is converted to a curve which avoids previously wrapped fibers and other aforementioned winding problems. Intermediate orientation planes along these curves are used to define how the robot moves to and from control points. Reorientation is described by rotating and moving as minimally as possible to align each orientation plane with the winding base plane. The simulation was only used as a visual check of the robotic movements with no integrated software collision detection. The (KRL) robotic code was written directly as a text file from the robotic simulation tool, saved as a.src file and uploaded on the Kuka control panel (KCP).

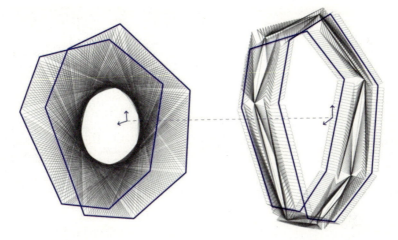

Fig. 8 Winding syntax polyline versus Robotic winding path

4 Results and Conclusion

An array of 35, highly differentiated components was fabricated through this research project. On average, components ranged from 0.5 to 3 m in diameter, had a structural height between 0.38 and 1.23 m and a winding time between 8 and 16 h. Depending on the component size and reinforcement, from 2 to 5 km of fiber rovings were used. Each lightweight component can be carried and lifted into place by 1–2 people. The complexity varied from 8 to 14 vertices per component and was up to 0.5 m out of plane (Fig. 2).

It was shown that highly differentiated fiber composite building elements can be manufactured using robotic coreless filament winding techniques. This novel fabrication process expands the possibilities of FRP to be used in architectural applications and lightweight structures. This method of fabrication comes with many challenges as well. Since the coreless winding process depends on fiber–fiber interaction to create the overall form, the overall morphologies are constantly adjusting as new fibers are wrapped. Further development in robotic control techniques such as real time robotic sensing could allow online adaptation of the robotic path. In this way the fabrication system could be an adaptive knowledge based fabrication process or even a behavioral materialization process where the final outcome is unscripted and a result of the process itself, rather than based on instructions for pre-determined geometry. The full implications of these novel design and fabrication processes are only starting to be explored but they open up a variety of future potentials and research opportunities in architectural applications (Fig. 9).

Fig. 9 Component details and 12 axis robotic setup

Future development towards a fully integrated building system would require many areas of investigation not explored within this research project, e.g. weatherproof enclosures and thermal insulation for complex geometries. This would raise additional questions regarding the respective materials and fabrication processes. Further investigations could open up possibilities for a higher level of functional integration which not only increases structural performance but includes capacities for climate modulating geometries, i.e. airflow guides for natural ventilation or self shading elements. The transfer of principles from biological role models of functionally integrated fiber-based material systems could inform alternative fabrication processes such as fibrous enclosure networks as formwork or adaptable pneumatic molds where, in both cases, the initial formwork could remain as a functional part of the FRP building component.

References

Anderson C (2012) Makers, the new industrial revolution. Random House, London, pp 185–191

Fischer J-R (2012) Optimizing digital organic freeform modelling for fabrication by using parameterization with glass fibre reinforced plastics (GRP) a case study. In: Henri A, Jiri P, Jaroslav H, Dana M (eds) Digital physicality—proceedings of the 30th eCAADe conference, vol 2, Prague, pp 181–190

Gang J (2010) The cook, the prospector, the nomad and their architect. In: Ruby I, Ruby A (eds) Re-inventing construction. Ruby Press, Berlin, pp 163–174

Gustafson P (2010) North sails 3Di. http://www.blur.se/2010/04/26/north-sails-3di-english-version/. Last accessed: 04.12.2013 6:38 GMT

Reichert S, Tobias S, Riccardo La M, Frederic W, Jan K, Achim M (2014) Fibrous structures: an integrative approach to design computation, simulation and fabrication for lightweight, glass and carbon fibre composite structures in architecture based on biomimetic design principles. CAD—steering of form. Forth-coming publication

Stage One (2009) "Case study: chanel—exhibition pavilion, London. http://www.stageone.co.uk/projects/chanels-mobile-art-container

Voigt P (2007) Die Pionierphase Des Bauens Mit Glasfaserverstärkten Kunststoffen (GFK) 1942 Bis 1980. Bauhaus-Universität Weimar. http://e-pub.uni-weimar.de/opus4/frontdoor/index/index/docId/821

Adaptive Part Variation

A Near Real-Time Approach to Construction Tolerances

Lauren Vasey, Iain Maxwell and Dave Pigram

Abstract This chapter introduces the concept of Adaptive Part Variation (APV) as a method where robotically automated fabrication and construction processes employ sensors and feedback to make real-time corrections to material and assembly processes by varying the geometry and location of future parts to respond to deviations between digitally defined and physically accumulating form. The potential disciplinary implications of the method are described followed by a comparison to existing approaches to providing tolerance for dimension error in architecture. As a case study, the material system of cold bending steel rod is utilized to investigate strategies for implementing Adaptive Part Variation within a fabrication workflow that includes the production, handling, and assembly of uniquely bent parts through synchronized robotic tasks and iterative sensor feedback. Two computer vision systems are compared to assess their value for APV processes. Finally, potential shifts in the deployment of procedural design methodologies are discussed in relation to adaptive automated construction processes.

Keywords Robotic fabrication · Formation embedded design · Digital fabrication · Robotic manipulation · Computer vision systems · Construction tolerances · File-to-factory

L. Vasey (✉)
University of Michigan, Ann Arbor, USA
e-mail: lvasey@umich.edu

I. Maxwell · D. Pigram
University of Technology, Sydney, Australia
e-mail: iain.maxwell@uts.edu.au

D. Pigram
e-mail: david.pigram@uts.edu.au

W. McGee and M. Ponce de Leon (eds.), *Robotic Fabrication in Architecture,*
Art and Design 2014, DOI: 10.1007/978-3-319-04663-1_20,
© Springer International Publishing Switzerland 2014

1 Introduction

The unstructured environment of the construction site together with the many factors responsible for significant dimensional error necessitates environmental awareness and the capacity to adapt to variation for the successful creation of architectural assemblies. The ability to work with multiple robots, customized tools and sensor inputs within architectural production provides a novel approach to address such challenges via automated and adaptive fabrication techniques incorporating feedback via systems of computer vision.

Robotically automated workflows bring into question many aspects of the relationship between design and production including the typical approaches to providing tolerance for irregularity. The concept of Adaptive Part Variation (APV) is an extension of craft based practices where parts are created or modified during sequential assembly processes to fit the accumulating scenario. This process is particularly well suited to the inherent variability of CNC fabrication in combination with sensor measurement, computational design and file-to-factory workflows.

The material system of cold bending steel rod is utilized as a case study to investigate strategies for implementing Adaptive Part Variation within a fabrication workflow that includes the production, handling, and assembly of uniquely bent parts through synchronized robotic tasks and iterative sensor feedback.

APV ultimately places priority on the processes and logics that govern an artifact's materialization, rather than the geometric specification of the artifact itself.

1.1 Automation

The ability to automate multiple custom and non-repetitive processes differentiates industrial robots from single-process computer numeric control (CNC) machines. However the majority of the existing industrial robots are still employed to complete repetitive production-line tasks. While there have been many attempts to bring robots into construction, most notably in Japan, these attempts have not gained significant traction. The research shift in recent years has been towards software innovations. This is true in the Construction and Automated Robotics (CAR) field (Balaguer and Abderrahim 2008) and it is true of the adoption of robotics research by architects. Significant to the architect's effort is an attempt to open the processes of automation to incorporate feedback loops between design and making. To support this ambition there have been numerous platforms created that skip dedicated CAM software (and its translation steps) and combine the capacity for algorithmic design and robotic instruction code generation within single CAD programs (Pigram and McGee 2011). While most were initially created without sensor feedback as a priority, the combined capacity for geometric production (and variation) and robotic control is a necessary predecessor setting the foundation for the next stage of evolution where computer vision informs production.

1.2 Disciplinary Concerns: Design Intelligence

The pursuit of automated workflows is not without disciplinary consequence. The field of digital fabrication has been lauded by many (the authors included) as allowing the Architect to take back possession of the production of the architectural artefact. The significance of this reclamation is to increase the number and quality of feedback loops between design and realisation. Conversely, if production intelligence is encapsulated within software in a standardised and black-boxed fashion—denying the architect the ability intervene—reciprocity between architecture and construction may be diminished rather than enhanced.

1.3 The Tectonics of Tolerance

All assembly processes demand attention be paid to dimensional variation and the provision of tolerances. This is particularly true within architecture where: parts are either large, come in high populations, or both; cumulative errors are significant; and site irregularities are common. There are at least three common approaches to tolerances in architecture. The first is to produce parts and assemblies of such high accuracy that cumulative errors are minimised to a level where they are acceptable. Much of the work typically labelled 'digital fabrication' follows this approach and it can be successful for autonomous material systems, small pavilions, sub-assemblies and other constructions where the interface between multiple trades is limited or non-existent. This approach does not scale well. At some point within the vast majority of architectural constructions one of the remaining two approaches is required to mediate more dimensionally precise sub-assembly/s with less. The second approach is to incorporate design details that accommodate potential dimensional errors such as oversized or slotted holes, shims, flexible connections, or joints that are welded or cast in situ. This is by far the most prevalent approach in construction. The third approach is commonly followed in an artisanal setting. Parts are crafted sequentially—or prefabricated parts are modified—to fit the accumulating situation. If there are dimensional errors, subsequent parts are adapted to follow the new course or to steer the construction back inline with the originally intended geometry. The creation and use of relatively inexpensive templates to test fit to site, substructure or already installed parts prior to the creation of the final part is a common example. These practices are still prevalent within construction in the work of joiners, carpenters, stonemasons and plasterers to name a few. The addition of sensors and feedback to the already growing use of robotic automation make extending this latter approach both a viable and attractive option within architectural fabrication and construction. We have named this extended approach Adaptive Part Variation.

2 Adaptive Part Variation

Feedback enabled digital production methods capable of fluently uniting part generation (design), file-to-factory protocols (communication) and tooling and sensing (making) permit a real opportunity to vary parts at the moment of manufacture. This is not to be confused with 'just- in-time' manufacturing where production timing—rather than geometry—is varied to reduce inventory holding costs. The speculation posited here is that APV modes of production present architecture with a powerful alternative to not only the mass-standardisation of building components, a benefit common to all CNC methods, but also a novel approach to handling cumulative construction errors and with it an expanded conception of architectural tectonics.

An effective APV production chain consists of four intrinsically related elements: an automated fabrication and assembly method capable of supporting geometric variation (Sect. 3); an environmental 'vision' system to interrogate part geometry and cumulative assembly status (Sect. 4); a communications feedback loop that connects the part-geometry generation with the environmental vision system (Sect. 5); design protocols that define the possible responses to gathered environmental information (Sect. 6).

3 Case Study: Robotic Bending, Assembly and Welding

The tectonic of rod bending presented, forms part of a trajectory of fabrication research undertaken at the University of Michigan beginning in 2009 (Fig. 1). The material system—essentially a continuous line punctuated by bends in space—is geometrically open-ended owing to the incremental operation of bending and the ability to easily vary the distance, angle, and orientation between each bend operation. The process is an adaptation of well-established bending strategies used in manufacturing. The significance of this application is the development of the tool and process coupled with a generic platform—the robotic manipulator—that remains open to the designer (Fig. 2).

In order to begin to test the potentials of Adaptive Part Variation an experimental workcell was established. Two Kuka KR120HA 6-Axis industrial robots were connected via ROBOTEAM, Kuka's application for synchronising the motion of multiple robots in a single workcell. The first robot uses a collet-gripper to tend the floor-mounted bender with integrated shear with each controlled as an external axis. A laser range-finding sensor, mounted to the end effector of this

Fig. 1 Bending projects produced at the A. Alfred Taubman College of Architecture and Urban Planning, University of Michigan. From *Left* 'Bent' by Nick Rebeck and Kendra Byrne; 'Wave Pavilion' by Parke McDowell and Diana Tomova; 'The Clouds of Venice' by *supermanoeuvre* and Wes McGee

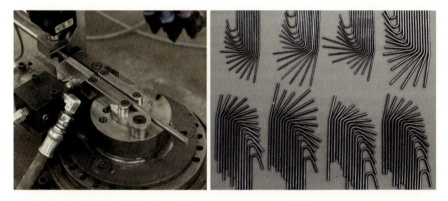

Fig. 2 *Left* Custom free-standing hydraulic bender and shear tended by a 6-Axis industrial robot, controlled as external axes. The die is able to rotate both clockwise and counter-clockwise. *Right* A physical catalogue of bend outcomes for rods of various thicknesses used to empirically calculate spring-back values

robot, serves as the primary means by which the system gathers necessary information about its environment, allowing for a feedback between an assembly and APV. The second robot has an end effector with an automated MIG welding torch (Fig. 3 left).

Together, the robots work collaboratively to bend, cut, position, and weld units into a larger assembly. The first robot undertakes all base functions necessary to produce each part: feeding, and actuating rod shearing and bending. Once the bend sequence is complete, the bender's die-gripper holds the bent component, allowing the first robot to re-grip and then reposition the part in space before the second robot welds it in place (Fig. 3 right). After each rod is bent, placed, and welded, a routine is launched to measure the aggregate assembly with laser scanner. The geometric information thus ascertained is then fed back to the CAD software

Fig. 3 *Left* Two Kuka KR120HA 6-Axis industrial robots with free-standing bender and shear, laser range-finder and MIG welding torch form a collaborative workcell for testing Adaptive Part Variation on steel rod assemblies. *Right* Welding in process

allowing for an update of the digital model. The geometry of all dependent rods is then re-computed based on known geometric relationships to those rods that have already been produced and assembled. Once this is accomplished the next rod in the construction sequence is produced. This process of scanning and part correction repeats until all parts have been produced and the assemblage is complete.

4 Vision Systems

If robots are to expand their operation beyond the structured environments of manufacturing where they are predominantly employed—typically moving only via explicit world-based coordinate positions–they must develop the ability to view and respond to changes in their environment. For this reason computer vision is becoming an expanding area of interest to the construction industry. Various approaches are being used to track a construction project over time (Bhatla et al. 2012) as well as to measure the tolerance between the *idealized* and *actualized* versions of a construction project (Tang et al. 2010). Such techniques are also being applied in the field of robotics to monitor parts as they undergo work. One example is the use of vision and sensory input coupled to robotic welders to monitor and respond to deformation (due to heat) throughout the welding process.

In order to establish an environmental awareness of the production pipeline, we must obtain critical environmental information as it unfolds. The adoption of multi-dimensional point-cloud scanning technologies is increasingly servicing such a role. In recent years there has been a surge in the number of off-the-shelf software packages capable of doing so as is the range of camera tools including the Kinect platform. These tools and applications primarily aim at very inclusive spatial models that can pose significant challenges to the realisation of efficient construction workflows. They simply generate excessive data much of which is extrinsic to the task at hand. The result being that computational expense is incurred searching the scanned model for relevant information and remapping it back to critical points in the input model.

To address the overabundance of information problem two alternate approaches to computer vision are explored below; each uses a lower (than 3-) dimensional scanning sensor and a measuring technique relevant to the specific geometry of the case study.

4.1 Computer Vision: Approach 01

The first approach works by identifying lines within photographic images taken from two or more locations. The hypothetical intersection of two adjacent polyline segments is a single point in any image. Because each point in an image defines a line from the camera position to that point, two images taken from known positions are sufficient to three dimensionally locate identified points (as the intersection of the two discovered lines). By mounting two cameras onto two robots or moving a single robot mounted camera to take two images we can at all times precisely know both the position and direction each image is taken from and are therefore able to employ this technique. When completed serially this process allows us to map the coordinates of a polyline in space (Fig. 4). This method relies on the accuracy of mapping camera positions to real world coordinates and is susceptible to lighting and other influences typical to photographic methods. It also obviously relies on the ability to correctly identify lines within images, a task that becomes increasingly difficult as the population expands and when the emerging assembly conceals elements despite the ability to move each robot and therefore camera.

4.2 Computer Vision: Approach 02

The second computer vision method combines a laser range finder with a simple vocabulary of moves capitalising on the native capacities of robotic programming. Moving the rangefinder in a straight line that is perpendicular to its scanned path will yield a series of distance values. The local minimum of these values reveals when the axis of the sensor is in line with the centre of the rod. Because the position and direction of the robots flange are known, so too can be the position and direction of the attached rangefinder (via a transform stored as the tool definition) and therefore the returned distance value allows us to locate a single point on the outside of the rod. This information is not enough to rebuild the assembled geometry in its entirety. An equivalent second point is required. This is found by subsequently following a parallel scanning toolpath. The now two located points—

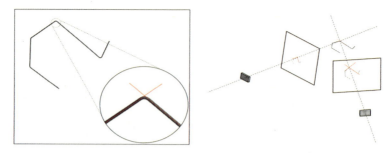

Fig. 4 *Left* A polyline corner identified in a single image. *Right* Two cameras, their picture planes, and the lines defined between each camera and the same polyline corner identified in each image. The intersection defines the corner in 3-dimensional space

each on the perimeter of the rod—reveal the vector direction of the rod in space (Fig. 5). If the diameter of the rod is known, or if the measurement is repeated using a scanning path out of plane with the first two, the location of the central axis of the rod can then be solved. This technique allows the precise definition of any polyline segment in world coordinates.

4.3 Computer Vision: Findings

Both the dual-photo and the laser rangefinder methods allow for the accurate locating of rods within space while reducing the amount of data required to do so when compared to 3-dimensional point-cloud scanning techniques.

Of the two methods, the second is distinctly superior, replacing the complexity of recognising details within images with simple numerical analysis. The laser rangefinder method allows for the complete reproduction of the assembled geometry into the digital environment with a minimal transference of information. Essentially, the robot returns only two points for any line segment. Additionally, geometric processing can be done directly in KRL machine language. This technique clearly demonstrates that rather than aiming at an exhaustive model of reality, a more abstract yet specific and efficient technique for achieving necessary feedback is preferable.

5 A Feedback Loop Framework

Inline communication between robot and the digital environment is enabled through the software 'KUKA.RobotSensorInterface', which allows continuous communication via an Ethernet connection. Initial processing of sensor data is computed in the robot controller utilising KUKA Robotic Language (KRL).

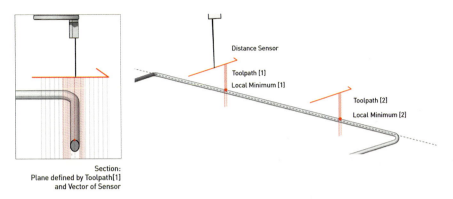

Fig. 5 Localization technique using parallel scanning toolpaths and a laser rangefinder

Subsequent computational processing of the 3-Dimensional data thus acquired, including updating the digital model and calculating dependencies, determining contingent responses and regenerating instruction code is completed with the CAD platform '*Rhinoceros 3D*' employing Python algorithms including the set of modules developed by the authors together with Wes McGee, Mark Meier and others under the name 'SuperMatterTools' (Fig. 6).

6 Designing Protocols of Response

In contrast to the static and predetermined nature of conventional construction details that must consistently cater to the worst-case, Adaptive Part Variation opens the possibility of deploying distinct conditional responses to various degrees of dimensional variance. A notable example of such response is *In-situ Robotic Fabrication*, conducted at the Swiss Federal Institute of Technology (ETH) by Gramazio and Kohler, Chair for Architecture and Digital Fabrication. This project implemented live scanning to allow a robot to recalibrate its position and make near-real-time adjustments during the construction process, a critical step towards making robots on the construction site a more viable reality (Helm et al. 2013). Their subsequent project 'Stratifications'—exhibited at The Bartlett, London as part of Fabricate 2011—employed scanning together with conditional part selection (as opposed to real-time part variation) and adaptive placement protocols. Conditional part selection was deployed to correct accumulating wall alignment errors. Simply, one of a given set of possible brick primitives featuring various degrees of non-parallel beds were simply substituted for an assumed flat brick to maintain wall verticality. The more complex adaptive placement protocols allowed the system to follow coursing irregularities and even to negotiate rogue bricks deliberately added during construction by humans demonstrating the capacity to cater to the unexpected. Critically the outcome produced may contain significant geometric departures from the originating digital model.

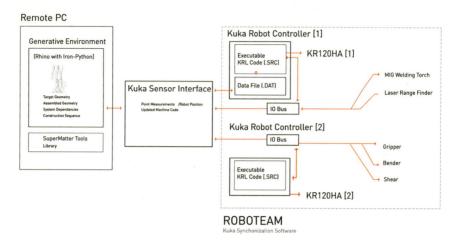

Fig. 6 System communication

The examples above point towards two key shifts. First, a necessary expansion of design beyond the absolute dimensional specification of artefacts towards the definition of relational systems that underwrite design intent whilst accommodating permitted forms of variance. To paraphrase Bernard Tschumi: "rather than condition the architecture; set the conditions for architecture." Second, that explicit positions, the primary information essential to all machine code, will ultimately be superseded by contingent positions adaptively defined based on designed relationships between constituent elements: their tangencies, adjacencies, and structural dependencies (Fig. 7).

Immediate considerations may include the rate at which dimensional variations are corrected. An immediate dimensional correction may result in an obvious compositional "glitch". A slower recalibration, or serial adjustment over a suite of proceeding parts alternately offers a viable approach to the concealment of errors, or an opportunity to embed additional systemic qualities within the work (character). Here the 'grain' of correction may begin to celebrate the techniques employed, or the material variances found, in a manner not dissimilar to the traces left by traditional craft practices. The software paradigm of control code crucially enables such behaviours to be tuned, developed, and ultimately embedded within the creative process itself.

7 Conclusion

Robotic fabrication in combination with algorithmic design methodologies and file-to-factory workflows affords new pathways of reciprocity between design and fabrication. The addition of computer vision sensors extends this paradigm adding a further feedback loop to include responses to real-time events and dimensional errors prevalent during construction.

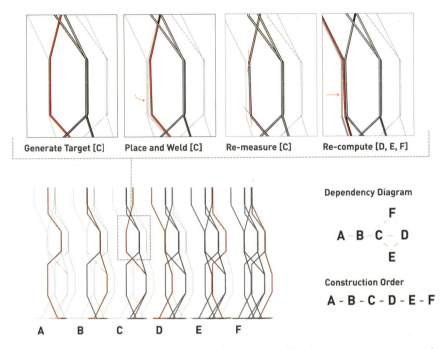

Fig. 7 Sample iteration of workflow: Each rod is represented by its bend sequence, geometric properties, and dependencies. Dependent rods are re-computed before the next rod is bent, placed, and welded and the process is repeated

Adaptive Part Variation is a strategy where the real-time redefinition and fabrication of parts occurs during the actual process of assembly thereby allowing detected errors to trigger a conditional design response. Necessary preconditions to establishing APV workflows are: the ability to progressively measure an assembly in process; to recognise key features within that assembly; and to update the digital model to register variance. The two computer vision approaches outlined here reveal that minimal and targeted approaches to information gathering have advantages over more exhaustive approaches and can leverage the robots inherent location awareness and specifics of the process geometry to gain further efficiencies.

Adaptive Part Variation demands a shift in the positioning of design intent away from static geometric descriptions towards relational frameworks and the definition of contingent responses to error. One immediate consequence of this shift is a radical reappraisal of prevailing approaches to the provision of construction tolerances. Typically, architectural details must consistently cater to a predicted worst-case, APV however, affords the definition of an enriched set of design responses that can directly address each actually occurring situation.

Adaptive Part Variation suggests that the pursuit of automation can be framed as the search for a new form of design intelligence, where the logics of design and making intermingle within a reflective design-to-construction model.

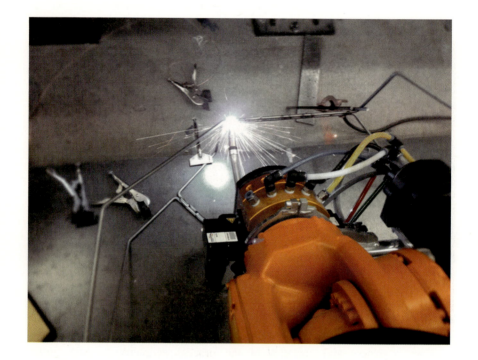

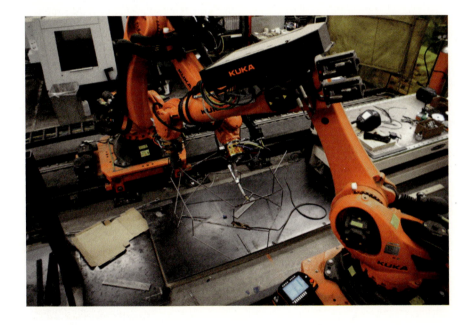

Acknowledgments The authors are incredibly grateful to Dean Monica Ponce de Leon, for establishing and continually supporting robotics research in the FABLab at the University of Michigan as well to FABLab Director Wes McGee for his continued collaboration. In addition, to FABLab Coordinator Aaron Willette for assistance and input with the paper. Significant feasibility studies were conducted by current students at the University of Michigan, Zac Potts and Andrew Pries.

References

Balaguer C, Abderrahim M (2008) Trends in robotics and automation in construction, IN-TECH, pp 1–20

Bhatla A, Choe S, Fierro O, Leite F (2012) Evaluation of accuracy of as built 3d modeling from photos taken by handheld digital cameras. Autom Constr 28:116–127

Helm V, Willmann J, Gramazio F, Kohler M (2013) In-situ robotic fabrication: advanced digital manufacturing beyond the laboratory. Gearing up and accelerating: cross-fertilization between academic and industrial robotics research in Europe. technology transfer experiments from the ECHORD project. Springer, Berlin, pp 63–85

Pigram D, McGee W (2011) Formation Embedded Design: A methodology for the integration of fabrication constraints into architectural design. In: ACADIA 11: integration through computation. Proceedings of the 31st annual conference of the association for computer aided design in architecture (Banff: ACADIA), pp 122–131

Tang P, Huber D, Akinci B, Lipman R, Lytle A (2010) Automatic reconstruction of as-built building information models from laser-scanned point clouds: a review of related techniques. Autom Constr 19:829–843

All Bent Out…

Adaptive Fabrication of Bent Wood Assembles

**Thibault Schwartz, Joshua Bard, Madeline Ganon,
Zack Jacobson-Weaver, Michael Jeffers and Richard Tursky**

Abstract All Bent Out… explores adaptive fabrication techniques for robotically constructing steam-bent wood assemblies. The following paper discusses a constellation of hardware and software tools that leverage the material constraints of steam bending (e.g. spring back, irregular grain) as opportunities to develop adaptive fabrication workflows where predetermined machine tasks can be informed by sensor-based events. In particular, research in coordinated motion control of two industrial robots and environmental sensing help to negotiate discrepancies between intended, digitally modeled geometries and actual, physically bent wood assemblies.

Keywords Adaptive fabrication · Real-time sensing · Computer vision · Steam bending

T. Schwartz (✉)
HAL Robot Programming & Control, Argenteuil, France
e-mail: ts@thibaultschwartz.com

J. Bard · M. Ganon · Z. Jacobson-Weaver · M. Jeffers
Carnegie Mellon University, Pittsburgh, PA, USA
e-mail: jdbard@cmu.edu

M. Ganon
e-mail: mgannon@andrew.cmu.edu

Z. Jacobson-Weaver
e-mail: zjacobso@andrew.cmu.edu

M. Jeffers
e-mail: mjeffers@andrew.cmu.edu

R. Tursky
Graduate School of Design, Harvard University, Cambridge, MA, USA
e-mail: richardtursky@gmail.com

W. McGee and M. Ponce de Leon (eds.), *Robotic Fabrication in Architecture,
Art and Design 2014*, DOI: 10.1007/978-3-319-04663-1_21,
© Springer International Publishing Switzerland 2014

1 Motivation

All Bent Out... reconsiders the traditional fabrication technique of steam bending natural hardwoods—a process pioneered by Michael Thonet (1796–1871)—using contemporary robotic technology. The focus of the workshop resides in a combined interest in material deformation, in particular the difficulties present in predictably free-form-bending natural hardwoods, and in the robotic assembly of bent components. To address material constraints in the process of bending and assembly, the workshop will develop an adaptive fabrication workflow, drawing from recent research by the authors in coordinated motion control of two industrial robots and real-time sensing. Adaptive fabrication is a responsive construction approach that allows a task to update based on data received from external sensors and events. This approach enables one to negotiate the translation from intended (digital) to actual (physical) material assemblies of bent wood.

2 Challenges Present in Bent Wood Assemblies

Steam bending is a traditional wood-working technique often used to construct furniture and other curvilinear industrial objects. As a proto-industrial technique, popularized by the iconic No. 14 Café Chair of Michael Thonet, steam bending has long captured the collective imagination of designers across many disciplines. The technique achieves curved material deformation in air-dried wood without the use of toxic adhesives. During the process, heat and moisture penetrate the wood's fibers and lower the elastic limit of the material. Bending stresses are introduced (beyond the elastic limit) resulting in a permanent set of the material and the resulting curved form (Hoadley 2000).

The constraints of applying robotic techniques to bent wood assemblies are a product of wood's unique material behavior during the process of steam bending. Many of the difficulties of free-form bending derive from the fact that wood is not a uniform material: "Steam bending has shortcomings. The most troublesome is accurately predicting springback... In steam bending the results depend upon the grain structure of each piece of wood. Local eccentricities—knots, checks, and cross grain—will affect the final curve... This disadvantage becomes critical when exact duplicates must be made" (Keyser 1985). Small changes in grain (e.g. runout or knots) can significantly alter the shape of a bend or sometimes result in material failure and careful handling/drying after bending is needed to minimize springback.

In addition, the nature and direction of forces applied during bending are important due to wood's anisotropic qualities. Often the limiting factor leading to material failure in bending is the introduction of axial tension on the outer, convex side of a bend. While steamed wood can compress up to 30 % of its original length it can elongate by a maximum of 2 % (Hoadley 2000). Industrial processes often deploy expensive, dedicated molds for repetitive parts to mitigate tension failure during bending. A steel tension strap can also be used but this often constrains bends to two dimensions and precludes free-form bending. One such example is the beautifully designed Timber Seasoning Shed from the Architectural Association's Design and Make graduate program where dimensionally unique parts are produced with a variable, table top formwork to form an undulating lattice shell (designandmake.aaschool.ac.uk/timber-seasoning-shelter/).

Wood's combined heterogeneous qualities introduce issues of unpredictable geometric translation from digitally modeled curvature to physical bending with repeatable robotic motion. This geometric error compounds when positioning multiple components into larger assemblies. Thus, despite the sub-millimeter accuracy of industrial robots, these combined factors lead to fabrication error beyond acceptable tolerances.

This workshop explores the potential to bend non-repetitive, three dimensional parts and aggregate them directly into architectural assemblies using robotic workflows. To achieve these aims wood's material behavior during steam bending can be addressed through adaptive fabrication techniques where bending irregularities can be analyzed during fabrication and digital models can be updated to respond in real-time. The following sections will discuss relevant precedent and the development of robotic software and sensing tools aimed toward this end.

3 Previous Work

A number of recent projects exploring bent architectural assemblies illustrate the problem space the workshop seeks to investigate. In particular, the workshop is focused on the deformation of heterogeneous materials and the robotic assembly of volumetric framing systems.

Case Study I: Spring Back, a steam-bent gateway structure, is one such example where CNC fabrication and parametric modeling were used to generate geometrically unique bent wood components (Bard et al. 2012). Precise registration holes were cut near each member's endpoint with a CNC water jet cutter. These registration holes located components in a variable bending formwork and later became connection nodes in the final structure.

In this case, the problem of modeling the irregular behavior of bent wood was not solved; rather, the influence of compound error in the assembly was mitigated by making part-to-part connections at key indexed locations. While this approach did allow for the efficient production of geometrically unique bent wood components, one drawback of this approach was that materially intensive falsework was needed to position components to erect the final structure. Large plywood forms were cut to bundle components together in three sub-assemblies that were then erected using a separate set of scaffolding (Fig. 1).

Case Study II: High Wire a project developed during a robotic fabrication seminar at the University of Michigan in 2012 addressed issues of robotic

Fig. 1 *Left* water jet cut blank with registration holes, *center* materially intensive formwork for component assembly, *right* spring back installed

Fig. 2 High wire assembly using robotic formwork and final installation

assembly in bent steel rod frames. The project incorporated robotic bending workflows previously developed by research teams at the University of Michigan in projects such as the "Clouds of Venice" installation at the 2012 Venice Biennale (Pigram et al. 2012). One aspect of the bending workflow extended through the seminar was the development of re-positionable, robotic falsework for the assembly of a large number of uniquely bent members into a steel space frame. Students developed a mountable steel fixture with interchangeable templates that was robotically positioned to weld individual bent rods into larger sub-assemblies (Fig. 2).

This strategy enabled the fabrication of an inhabitable (3 m × 8 m × 3 m) volumetric assembly without the use of extra falsework during installation. Although High Wire shares an affinity for material deformation of linear members and the assembly of structural frames the project remains distinct from bent wood assemblies due to the differing material behavior of wood's anisotropic and steel's isotropic properties. As a result, falsework for welding modules could be positioned directly from offline data with little need to check physically bent pieces for fidelity to the intended assembly.

Fig. 3 HAL software architecture, in relation to external sensors and robot controllers

4 Software Development for Adaptive Fabrication Using HAL

4.1 Development Context

Adaptive control, as a technique for the manipulation of a predetermined machine task informed by external sensors and events, is a key topic of software research. Notwithstanding substantial development in many areas of robotics and computer science, adaptive control is still underutilized in the domains of architecture and construction, despite many promising applications (e.g. automated compensation of an existing part, real-time adaptation to deformable material behavior, interaction with dynamic environments, etc.). The following sections focus on the implementation strategies for HAL (Schwartz 2012) users to integrate feedback mechanisms—and thus adaptive control—into ABB robot programs (Fig. 3).

4.2 Adaptive Control Scenarios

In order to understand the implementation logic we will depict, it is necessary to analyze several usage scenarios:

Scenario 1: The user wants to modify a single position, relative to a tool or reference axis system calibration procedure. In this case, the compensated position can directly be linked to a single frame variable, and will automatically impact any further movement instruction without additional computing efforts. An alternative is to use multiple search routines (using *SearchL* for example) to obtain an equivalent compensation by merging multiple local corrections (along any direction).

Scenario 2: The user wants to modify a single position in a set, in order to dynamically compensate for an imprecision detected during the execution of a toolpath. The compensation will only affect this position and will necessitate an update of its coordinates in the code. It will require an interruption of the execution in order to take the compensation into account every time a modification is required.

Scenario 3: The user wants to globally refine the positioning of a toolpath. The ABB correction generator feature can be used, and will allow the user to activate, deactivate, or update a global correction during the execution to make it local if necessary. Another solution is to modify the reference axis system of the toolpath and/or the active tool, by adjusting their declaration by the rotation and translation that needs to be compensated (Scenario 1). A third solution, involving more computation, is to apply compensation via a displacement frame. For this scenario, the compensation can be executed iteratively. These scenarios can then be merged and tweaked to create some automated toolpath teaching procedures, step-by-step compensated manufacturing processes, etc.

It is obvious that the organization of the communication routines channeling the compensation measures will have a strong impact during the execution of each of these scenarios. Multitasking is an option available in ABB controllers starting with the S4 versions, but requires some manual operations on the teach pendant unit and a reboot of the machine to create or a delete a parallel task. To eventually gain in performance on the communication latency, it could be useful to use a background task to constantly listen to communications with the computer, but due to the inability to administrate tasks automatically, the workshop will focus on a solution involving a single task, this solution allows one to switch instantaneously between traditional programs and interactive programs with minimal latency.

4.3 Implementation Logic

The implementation logic of adaptive tasks is handled in HAL by the automated inclusion of the programmed process/toolpath into a generic "feedback-friendly" RAPID module. This module embeds five different logic blocks (Fig. 4):

- Variable declarations, including positions, speeds, tools, work object and zone presets, and all the necessary variables to be accessed by the communication and correction routines.
- Main process/toolpath procedure, including the motion instructions. Depending on the selected compensation mechanism, one or multiple instructions can trigger the communication routines, using specific RAPID instructions (*TriggL*, *TriggJ*, *TriggIO*, etc.).
- Communication routines that can automatically select and parse the feedback values coming from an internal or external sensor. Different communication methods are available depending on the amount of data to be transferred: RMQ messages (for short instructions), TCP socket messages (for short or long instructions), or IO monitoring (for analogue sensing or digital switches).

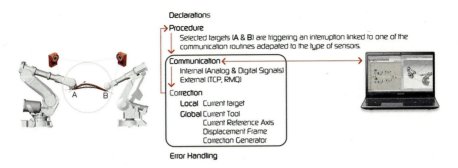

Fig. 4 HAL feedback implementation logic to be executed on the robot side

- Correction routines, applying compensation measures as explained in the list of scenarios. These correction routines can also be looped with the communication routines until an acceptable tolerance is reached.
- Error handling routines, allow one to continue the process or propose additional communication options on the teach pendant unit when a compensation operation or the communication with external devices failed.

4.4 Current Limitations and Further Development

While this approach of generic feedback loop integration has numerous benefits especially concerning the usage flexibility, it is true that a multi-threaded application could bring several improvements in certain scenarios. One main limitation of the current single threaded feedback comes from the temporary interruption of the initial process triggering the correction routines, as it disables smooth process during the compensation. If the communication and correction routines were executed in parallel tasks, it could diminish this small inactivity time, and eventually lead to better overall precision for intrinsically continuous applications. Another limitation comes from interruption and communication latency, which exceed real-time (<10 ms) latency requirements, even if acceptable response can be measured with this system (100 < 250 ms).

5 Sensing Tools for Adaptive Fabrication in Wood Assemblies

The following section describes open-source toolsets developed by research groups at Carnegie Mellon University. These toolsets combine techniques in proximity sensing, computer vision (CV), lo-fidelity force feedback, and motion capture (MOCAP) to augment standard industrial robot configurations with real-time control. Each custom tool encodes a contextual awareness of the immediate physical environment within the robot's work cell. When these tools are layered together, they demonstrate how live control of an industrial robot can be safely driven by environmental stimuli, to augment standard on-line programming. These toolsets can play an important role in making fabrication and assembly processes for robotic steam-bending more adaptive to the material constraints of natural wood.

5.1 Vision Capture

Computer vision provides environmental awareness and can generate motion paths from visual cues in a robot's immediate context. For example, *Stroke*—a program used to dynamically draw tool paths for an industrial robot on a physical 2D work surface—incorporates computer vision to capture and translate hand sketches into robot position targets. With robotic steam bending, these same techniques can be adapted for checking physical bends in wood members against an idealized digital model (Fig. 5). During the coordinated bending process, a camera can capture and extract a trace of the physical curvature of the deforming wood. Once captured, the deviation between a projection of the desired digital curvature and actual physical curvature can be used to adapt the robots' bending motion. Projecting the digital curvature onto the physical component offers another layer of visual error checking during bending.

5.2 Force Feedback

RoboMasseuse is a project that modifies a light payload industrial robot with a force feedback end-effector. Force feedback is an important component in adaptive fabrication, as it has the potential to detect *when* and *how* a robot engages a particular surface. Pressure sensors embedded within the therapeutic end-effector enable the robot to safely give back massages to its human operator. Streaming sensory data is used to map tactile feedback from the operator to initiate robot commands for pushing harder or softer against the user's back. This force feedback end-effector can be directly applied to control and coordinate the variable pressure needed to stress the wood during the robotic steam-bending process (Fig. 6). Whereas the forces controlling *RoboMasseuse* were based on human limits, *All Bent Out…* will modify robot movement based on measured forces approaching the minimum stress to failure point of the bent wood. Therefore, through simple recalibration, this force feedback tool can become a device for adaptive forming in robotic steam bending.

5.3 Motion Capture

Motion capture is a vision strategy for tracking and collecting long periods of position/orientation data of a subject or rigid body. For example, *Pose*, uses rigid body tracking to capture the motion of handheld tools for collaborative construction tasks. MOCAP Camera arrays positioned within a work cell can aid in the adaptive assembly of bent wood members (Fig. 7). Placing tracking markers

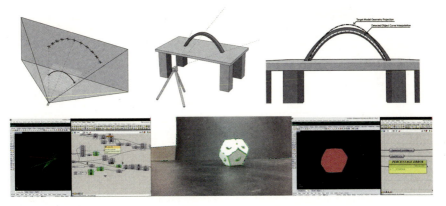

Fig. 5 *Left* digitally project desired curve. *Center* capture outline of physical bend. *Right* compare desired curve with physical bend and update motion

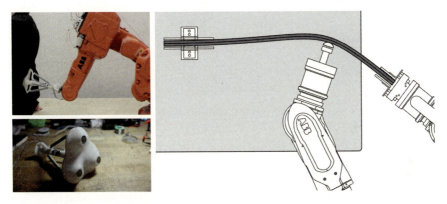

Fig. 6 Custom forming tool gauges pressure feedback from stressed wood

onto bent components and known registration points in the assembly area will enable each bent piece to be dynamically positioned into the overall configuration. After each component is placed in the assembly motion tracking positions can be checked against reference points in the digital model and be used to update the digital description of each successive component in the remaining assembly.

6 Workshop Outlook

All Bent Out... will stage three areas for material preparation and steam, material deformation, and assembly (Fig. 8). After wood is pre-soaked and steamed it will be moved to a coordinated bending cell within the overlapping work envelopes of an IRB 4400 and an IRB 6640. While in the bending cell the steamed wood's endpoints will be captured by a table gripper and the end of arm gripper of the IRB 6640. A third, force sensitive, end-effector on the IRB 4400 will engage the material at

Fig. 7 *Left* calibrating motion capture and robot path. *Right* markers placed on physical components map the deviation between digital model and physical assembly

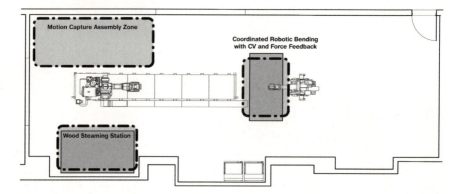

Fig. 8 Coordinated bending with an ABB IRB 4400, IRB 6640, and table fixtures. The IRB 6640 will position each component in the MOCAP assembly area

mid-span and deform the material into a desired shape. After each component is bent it will be checked with a digital camera and tagged with infrared reflectors. The component will then be repositioned along the 6640's 6 m linear axis in the MOCAP work cell and tracked in its final location. Tracking data will be compared against digital models of the assembly and discrepancies will be mitigated by generating new target positions for subsequent components. Participants of the workshop can anticipate testing this workflow through the fabrication of a bent wooden space frame with 30–50 members. Specific geometries will be refined through open-ended explorations in possible bent forms at the outset of the workshop.

7 Conclusion

The authors look forward to developing adaptive fabrication techniques in response to the unique challenges of robotically steam bending natural wood and anticipate that the synthesis of custom hardware and software tools developed for the workshop will extend the architectural designer's abilities to imaginatively utilize contemporary technology in fabrication settings. In particular the hope is that new opportunities to work with heterogeneous materials will drive further advancements in adaptive fabrication relative to the means and methods of constructing the built environment.

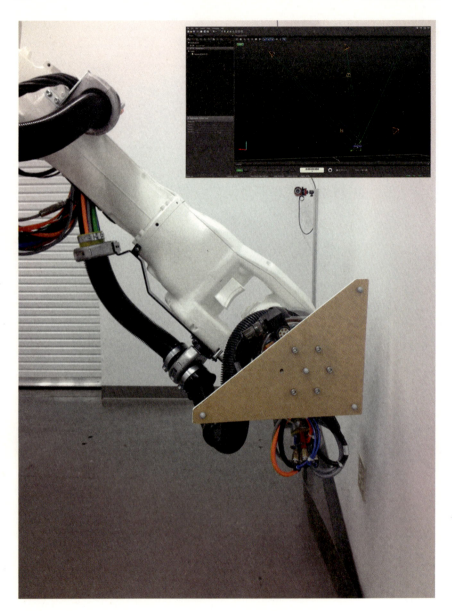

References

Bard J, Mankouche S, Schulte M (2012) Digital steam bending: re-casting thonet through digital techniques. In: Synthetic digital ecologies: proceedings of the 32nd annual conference of the association for computer aided design in architecture (ACADIA), San Francisco, California, pp 117–126

Hoadley R (2000) Understanding wood: a craftsman's guide to wood technology. The Taunton Press, Newtown

Keyser W (1985) Steambending: heat and moisture plasticize wood, fine wood working: on bending wood. The Taunton Press, Newtown, pp 16–21

Pigram D, Maxwell I, McGee W, Hagenhofer-Daniell B, Vasey L (2012) Protocols, pathways, and production. In: Proceedings of the Rob|Arch conference, Vienna, Austria, pp 143–147

Schwartz T (2012) HAL: extension of a visual programming language to support teaching and research on robotics applied to construction. In: Proceedings of the Rob|Arch conference, Vienna, Austria, pp 92–101

Design Approaches Through Augmented Materiality and Embodied Computation

Ryan Luke Johns, Axel Kilian and Nicholas Foley

Abstract With the increase of research experiments engaging the potential uses of industrial robotics in architecture, it becomes necessary to categorize the components of these exercises within a number of directions and motivations which can be related in the field, and to their larger consequences within the architectural discipline. In this chapter, we present a number of approaches to robotic design/fabrication exercises that deal with information, interactivity, and material dynamics. We outline the concept of 'informed operator' fabrication, in which computer numerical control (CNC) is used as a means for providing information to the operator in addition to the conventional use of providing instructions to the machine. Building upon this, the concepts of embodied computation and augmented materiality are discussed within the context of robotic manipulation. Embodied computation is introduced as enabling a protraction of the design/fabrication sequence beyond the scope of digitally controlled tools, such that robotic or human actions trigger ongoing material responses. Augmented materiality is presented as the human occupation and influence upon this "material in the loop" procedure, as enabled through interactive and digitally mediated interfaces.

Keywords Augmented materiality · Embodied computation · Procedural fabrication · Architectural robotics · Design workflows

R. L. Johns (✉) · N. Foley
Greyshed, Princeton, USA
e-mail: ryan@gshed.com

N. Foley
e-mail: nick@gshed.com

R. L. Johns · A. Kilian
Princeton University, Princeton, USA
e-mail: akilian@princeton.edu

W. McGee and M. Ponce de Leon (eds.), *Robotic Fabrication in Architecture, Art and Design 2014*, DOI: 10.1007/978-3-319-04663-1_22,
© Springer International Publishing Switzerland 2014

1 Introduction

In most models of increasing automation in production, robots are replacing the human performance of formerly manual tasks with increases in efficiency, speed and precision. Such a reading implies the isolation of the design process as an almost exclusively human cognitive process, which is manifested in physical form in connection with 3D design software and digital fabrication tools. The promise of such a separation and streamlined process is the removal of the intermediary segments from the process pipeline and the promise of more authority and control for the designer. However, this separation also removes the human from being involved in the tangible execution of the design. Designer intervention in the production process as it happens is one strand to be discussed here. Another is the possible role the physical material can take on if sensor feedback is included in the fabrication process. Once the material state is known, the progression of its manipulation can be updated in response to material processes that go beyond the immediate manipulation and are incorporated into the design. Thus, the process can become one of *embodied computation*. The extension of this concept through the application of augmented digital and material interfaces enables a form of *augmented materiality*.

With the rising quantity of experimental projects which engage architectural robotics, it becomes increasingly necessary to outline a range of methodologies and motivations by which these projects can be placed within a broader discourse. In this chapter, we outline a subset of approaches and areas of interest for the workshop which are presently practiced and explored at Princeton University's *Embodied Computation Lab* and by *Greyshed*. Rather than beginning with the framework of a specific physical procedure, the intent is to provide a conceptual and methodological scaffold upon which an array of procedures can be assembled. By establishing generalizable principles through a variety of research experiments, a *conceptual* "morphospace" (Menges 2012) is defined which provides a specific region of research to be navigated within the context of architectural robotics.

2 Informed Operator Fabrication

The rising popularity of industrial robotics in architecture runs in parallel to the increasingly expansive set of "open source and bespoken software applications" (McGee et al. 2012) which make these tools easier for designers to program. Such frameworks (as BootTheBot, HAL, KUKA|prc, Lobster, Mussel, Onix, PyRapid, Rhino2krl, superMatterTools, etc.) enable "highly informed" (Bonwetsch et al. 2006) operability with minimal development time for the end user. As high levels of complexity become easier to achieve, their intricacies can become more difficult to comprehend. While in some cases the fabrication process acts inherently as infographic (in example, the linear feed of a 3D printer or raster-based milling/

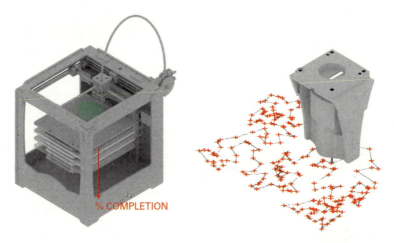

Fig. 1 The bed depth or build height in 3D printing acts as an infographic for its state of completion. In comparison, the correlation is unclear in robotic movements, i.e. drilling

etching/plotting serves as a physical progress bar for its completion), there is generally no clear corollary between the human operator's assumptions about the movement of a CNC machine and it's actual execution. Even with the simple addition of a path optimization algorithm, the machine's movements become significantly less predictable than with standard tabular Cartesian movement, and it becomes virtually impossible for the operator to know the position or order of subsequent moves (Fig. 1).

In order to satisfy the intent to unify the design and fabrication process "into an open system, where design decisions can be made while the physical manufacturing process is in progress" (Dörfler et al. 2012, p. 83), it is not only imperative that the designer can influence in-progress fabrication, but that he is capable of perceiving its peculiarities. By adding a system of callouts and interactive overlays which sync the physical actions of the fabrication process with the information of the digital model and its numerical control, the human operator becomes more informed as to the global significance of any singular machanic movement. As the absolute position of the robot and any robotically machined part are known by default, three-dimensional overlays can be achieved without necessarily requiring the complexity of computer-vision based tracking. Whether by digital projection, tablet interface, smartphone or AR-headset, guides which indicate information such as future toolpaths, positions of I/O triggers, registration marks, part-to-whole relationships, assembly instructions or points of possible interjection should evolve as an integrated byproduct of the machine code (Fig. 2). The most basic example of this is simply embedding messages into the robot code that appear on the control pendant, prompting action or requesting inputs. That which guides the machine should also guide the user.

Fig. 2 Toolpaths are displayed on the tablet and can be modified on the fly

3 Interaction

The integration of the design and fabrication process into a continuously inter-active workflow stems from the basic desire to reconsider the role of the human designer in the face of increasingly complex automation in fabrication. Rather than taking the stance of the Luddite by treating automation technologies solely as a shift away from human intuition and production, it is necessary to think about these elements as developing in parallel, co-dependently. The concern, however, is not without merit: a variety of robotic systems are quite simply existing vehicles with the human element removed from the loop: the earth digger or crane precedes the industrial robot, the drone and quadcopter are pilotless planes/helicopters, and the autonomous vehicle is a driverless car. In attempting to develop intelligent machines and operations which perform functions traditionally controlled by humans, we are encouraged to rethink our own role in these processes rather than completely severing the ties.

While the mechanical and numerical control of the robot might sometimes surpass that of the human, there are still numerous instances where the comput-erized system benefits greatly from the augmentation of human skills, such as image processing or spontaneous decision making (Branson et al. 2010; Willman et al. 2012; http://www.darpa.mil/NewsEvents/Releases/2012/09/18.aspx). By keeping the "human in the loop," the intuition and cognition of the operator augment the skills of the robot, just as the robot augments those of the designer. In

Fig. 3 *Left* **a** Human as noise generator: the robot's level of precision while drawing is linked to the real-time values of an EEG headset, such that the drawing becomes messier as the human loses focus. *Right* **b** [AR]chitecture: gesturally defining a loft surface and a brick wall simultaneously using the Kinect (Johns 2012)

such a loop, it is necessary to compress and "interlace" (Dörfler et al. 2012) the design and fabrication sequence such that there is a continuous exchange of information: essentially, operating on the scale of *Byte to Robot* rather than *File to Factory*.

In using interactive techniques to bridge "the divide between embodied input and embodied output," (Willis et al. 2010) we utilize human control and sensibility to provide a level of logical determinacy while simultaneously embracing the indeterminacy associated with improvisation. Interaction is therefore a means to augment machine logic while imbuing the artifact with the aura of manual manipulation and the proportionality of human gesture. This approach echoes that of Oskar Schlemmer and Johannes Itten, who emphasized the role of human intuition in design in the wake of the second industrial revolution as a balance between the Apollonian regulations associated with mechanical production and the free, rhythmic, Dionysian movement of the human form. As Schlemmer writes, "the initial impulse should be emotion, the stream of the unconscious, free, unfettered creation...If mathematical proportions and measurements are called in, they should function as a regulative." (Schlemmer, Oct. 1972)

Digital technologies have the potential of reuniting the architect with his a priori material intuition and design impulses while filtering those decisions through "regulative" computational tools which keep structural, proportional, or other coded guidelines in check. This is essentially the digital extrapolation of the idea that "all beautiful lines are drawn under mathematical laws organically transgressed." (Ruskin 1894) Such organic transgression can vary in influence, from simple noise generation (Fig. 3a), to gestural formation (Fig. 3b), or a combination of human input with natural or material phenomena (Fig. 6)—all of which benefit from an underlying algorithmic control. "Computers let us improvise better, with more notational density, with more variations possible in real

Fig. 4 Path followed by robot (*red*) versus material result of extruding silicon rubber (*blue*) into liquid soap

time, and with that particular merger of continuity and notation so difficult to achieve in material media." (McCullough 1996, p. 271)

4 Controlled Manipulation and Dynamic Components

The high degree of control provided by the robotic manipulator presents a means by which to experiment with more volatile unknowns such as human improvisation or material indeterminacy. In approaching robotic fabrication exercises, we generally focus upon one or several qualities, or robotic virtues, which lend themselves to the rethinking of traditional design/construction techniques through the assurance of certain stabilities. These properties are speed, strength, stamina, patience, precision, and synesthesia.

The predictability of these qualities serves to enable, among other things, procedural processes which engage dynamic, stochastic or embodied material properties. On example of such an approach is the "Procedural Landscapes" project of Gramazio and Kohler, which engages the material properties of sand (http://www.dfab.arch.ethz.ch/web/e/lehre/208.html). Another is our experimentation with "Buoyant Extrusion", in which complex geometries are created by synchronizing relatively simple robot movements with the extrusion of thermoset polymers into a similarly buoyant medium (Fig. 4).

5 Embodied Computation

In an extended framework, such material-based procedural experiments can be situated in the concept of embodied computation, and boundaries between design process and design artifact can be redrawn.

Embodied computation offers the possibility to shift part of the execution of formal manipulation from a top–down process of manipulating material in a static manner towards one where the robotic or human actions are only part of the operation and changes triggered by the action complete the process.

This opens up different forms of "Material in the Loop" possibilities. One is that of continuous dynamic interaction such as the robotic arm swigging a liquid in a mold such that the liquid distribution across the surface happens through the combination of gravity and centrifugal forces controlled by the robotic arm. The setup of a camera and the timed extraction of a single frame every half second allows for the visualization of a single, recurring liquid feature. This essentially demonstrates a simple design interaction with a liquid form generated through the embodied computation of the material and guided by a numerically controlled actuator (Fig. 5).

This raises interesting challenges for linking the digital state of the model with the physical. The simulation challenge here is not a precise predictive model but one that allows for constant synchronization between the physical and the digital state of the design process. The approach expands computational processes from digital processes being executed into physical form to include the stochastic behavior of material into the form giving process.

Embodied computation can be extended to connect the design process with the designed artifact itself. The work by D'Andrea's group at ETH Zurich shows an example where a computationally controlled quadcopter can recover algorithmically from a drastic physical change to its physical body (such as partial trimming of its rotor blades) through learning on the fly from the changed feedback it receives through its sensors (Mueller and D'Andrea 2011, 2012).

6 Augmented Materiality

6.1 Concept

The robotic manipulation of dynamic or stochastic materials has demonstrated the potential to result in novel constructs which take form through the application of the principles of embodied computation. Such constructs, while generally repeatable within certain tolerances, prove inherently difficult to generate based on specified design intent. As the resultant form does not develop as a direct parallel of a digital model, but rather as the result of a material reaction to a designed trigger, the outcome of such procedural experiments can be largely unpredictable

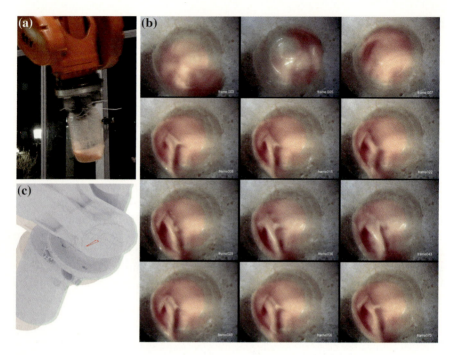

Fig. 5 Clockwise from *left* **a** Robotic swigging. **b** Fluid form reemerges every 7th frame. **c** Simple and repeating three-point-toolpath indicated in *red*

or unexpected. While the richness of the indeterminate material reaction is desired, it is necessary to explore means by which we may direct these reactions towards a result which more closely approximates the intent of the human designer. Through the use of visualization techniques which make the computer's algorithmic processes and material simulations apparent to the designer, and interactive tools which allow him or her to manipulate both the digital model and the material manifestation simultaneously, the result of such experiments can be more closely related to the embodied inputs associated with design intent. This type of workflow requires a quadripartite balance between the influence of the human designer, the robotic manipulator, the material properties, and the computer simulation, where no entity can operate without accounting for its impact upon the others. The concept of augmented materiality is understood as the encapsulating framework of such workflows. Essentially, it is a system in which interactive techniques enable the guided, real time manipulation of stochastic material systems—affording a degree of improvisation while maintaining the connection to the "highly informed" potential of digital models and tools. Augmented materiality can be understood as a means to imbue material craftsmanship with the qualities of digital fabrication such that algorithmic and robotic control act as additional material attributes. For the sake of this definition, we recognize craftsmanship as "simply workmanship using any kind of technique or apparatus, in which the quality of the result is not predetermined." (David Pie, cited McCullough 1996, p. 202)

Augmented materiality stems from the concept of "*Digital Materiality*", which "evolves through the interplay between digital and material processes in design and construction." (Gramazio and Kohler 2008, p. 7) Augmented materiality engages this interplay while focusing upon the human position in this dynamic: through sensory feedback and the physical overlay of digital information, it compresses the process of digital/material conversion into a feedback loop which supplements algorithmic and manipulative processes with human improvisation and intent.

6.2 Sample Application

The concept of augmented materiality is illustrated in the *Mixed Reality Modeling* project, which uses a robot-mounted heat gun (the material manipulator) to iteratively melt away material from a block of wax (the material). In this process, the robot is equipped with a 3D scanner and RGB camera (the sensors) which provide a colored point cloud of the physical materials and interface. The human user can indicate desired structural load forces on the wax by placing physical blocks. These are scanned and automatically placed in the corresponding location of the digital model. The software then proceeds to evaluate the structural necessity of each region of the wax block through topological optimization, accounting for the user-placed loads and support conditions. Following this calculation, the robot proceeds to heat and melt away areas that are the least structurally necessary. As the amount of wax that will melt away during each melt cycle, or the direction that it will flow and accrue is not precisely known, the process requires iterative scanning and recalculation. This iterative manipulation with the "material in the loop" simultaneously allows for "human in the loop" modifications: at any point in the process, the human can shift the loading conditions, indicate desired void areas by coloring on the wax, or physically modify the wax. Through the combination of user tracking and digital projection, the human operator is constantly informed as to the three dimensional calculations of the software, and the projected toolpaths and operations of the robot (Johns 2014) (Figs. 6, 7).

6.3 Generalization

In a generalized context, processes which engage augmented materiality must provide a means for embodied interaction from the human user (through the physical manipulations of objects or one of the increasingly large variety of intuitive human interface devices), and a means to inform the user as to the operations of the digital model and its physical manifestation (augmented reality).

Fig. 6 Prototypical augmented reality interface. Projection of digital information onto physical artifact

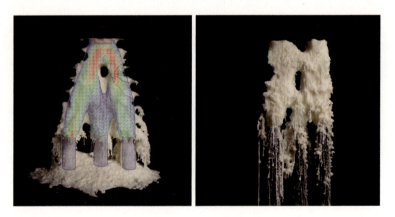

Fig. 7 *Left* topologically optimized digital model versus materially informed result. *Right* iteratively melted form with two loading points, three supports, and user indicated void

The physical process must inform the digital model through a network of sensors, and vice versa through a means of digitally controlled manipulation (robots or CNC devices of any form). This manipulation should not be entirely determinate in its effects, but according to the principles of embodied computation, should

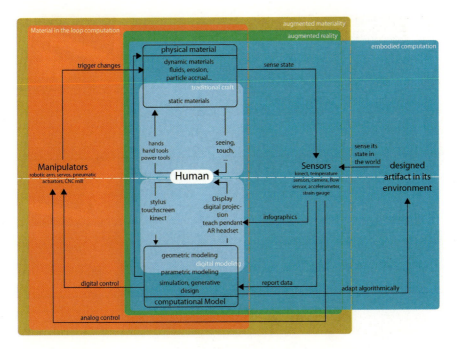

Fig. 8 Conceptual framework for augmented materiality and embodied computation

serve as a trigger for a more complex material reaction. The selected physical medium should therefore engage some level of indeterminacy or stochastic behavior, such as fluid dynamics, erosion, plant growth, animal behavior, etc. Thus, the designer is provided the capability to iteratively influence and craft stochastic systems while simultaneously benefiting from the informed control of computer algorithms which work to maintain coded parameters such as structural stability, volume, or program within established parameters (Fig. 8).

7 Conclusion

Recent work in robotic fabrication serves to augment the human with the precision, power, and speed of an automated process. The claim here is to expand this notion of augmentation to include the augmentation of the material that is being manipulated. The inclusion of sensing and feedback which report back on the state of the physical artifact as it is being changed allows for a closer fusion of human and robot actions. The concept of *embodied computation* is introduced as a protraction of the design sequence to include physical and material reactions which continue to occur after and in reaction to the specified trigger. *Augmented materiality* is then presented as the human occupation and influence upon such a cycle, as enabled through an interactive and digitally mediated interface.

References

Bonwetsch T, Gramazio F, Kohler M (2006) The informed wall: applying additive digital fabrication techniques on architecture. In: Proceedings of the 25th annual conference of the association for computer-aided design in architecture, Louisville, Kentucky, pp 489–495

Branson S, Wah C, Babenko B, Schroff F, Welinder P, Perona P, Belongie S (2010) Visual recognition with humans in the loop. European conference on computer vision (ECCV), Heraklion, Crete

Dörfler K, Rist F, Rust R (2012) Interlacing: an experimental approach to integrating digital and physical design methods. Rob|Arch 2012: robotic fabrication in architecture, art and industrial design. Springer, Vienna, 2013, pp 82–91

Gramazio F, Kohler M (2008) Digital materiality in architecture. Lars Müller Publishers, Baden, Switzerland

Johns R (2014) Augmented materiality: modelling with material indeterminacy. In: Gramazio F, Kohler M, Langenberg, S (eds) Fabricate. gta Verlag, Zurich, pp 216–223

Johns R (2012) Augmented reality and the fabrication of gestural form. Rob|Arch 2012: robotic fabrication in architecture, art and industrial design. Springer, Vienna, 2013, pp 248–255

McCullough M (1996) Abstracting craft: the practiced digital hand. The MIT Press, Massachusetts

McGee A, Feringa J, Søndergaard A (2012) Processes for an architecture of volume: robotic hot wire cutting. Rob|Arch 2012: robotic fabrication in architecture, art and industrial design. Springer, Vienna, 2013, pp 28–47

Menges A (2012) Morphospaces of robotic fabrication: from theoretical morphology to design computation and digital fabrication in architecture. Rob|Arch 2012: robotic fabrication in architecture, art and industrial design. Springer, Vienna, 2013, pp 28–47

Mueller MW, D'Andrea R (2011) Critical subsystem failure mitigation in an indoor UAV testbed. International conference on IEEE/RSJ, 2012

Mueller MW, D'Andrea R (2012) Critical subsystem failure mitigation in an indoor UAV testbed. IEEE/RSJ International conference on intelligent robots and systems. Vilamoura, Algarve, Portugal

Ruskin J (1894) The complete works: Fors Clavigera, vol 3. Bryan, Taylor & Company, New York, p 425

Schlemmer T (ed) (1972) The letters and Diaries of Oskar Schlemmer. Trans. Krishna Winston. Wesleyan University Press, Middletown, Connecticut

Willis K et al (2010) Interactive fabrication: new interfaces for digital fabrication. In: Proceedings of the fifth international conference on tangible, embedded, and embodied interaction. ACM, New York, USA, pp 69–72

Willman J, Gramazio F, Kohler M, Langenberg S (2012) Digital by material: envisioning an extended performative materiality in the digital age of architecture. Rob|Arch 2012: robotic fabrication in architecture, art and industrial design. Springer, Vienna, 2013, pp 12–27

Material Feedback in Robotic Production

Plastic and Elastic Formation of Materials for Reusable Mold-Making

Felix Raspall, Felix Amtsberg and Stefan Peters

Abstract The success of CAD/CAM in architecture relies on the consistency between geometric information, material processes, and physical results. However, when material processes exceed a level of imprecision, the correlation between intended geometry and physical output cannot be secured and, therefore, conventional workflows are inadequate. This research investigates the expansion of one-way design-to-fabrication processes (from digital to physical) through the addition of feedback control. It develops methods to adjust fabrication instructions while it is occurring, in order to guide imprecise production into useful outputs. Experiments using feedback control to produce clay molds and to adjust a universal formwork are discussed.

Keywords Robotic fabrication · Real-time feedback · Casting · On-line robotic control · Computer Vision

1 Introduction

The application of digital tools in the design industry, typically referred as computer-aided design and manufacturing, or CAD/CAM, has significantly advanced how designers create, analyze, modify and fabricate their projects. Within this framework, recent research on design-to-fabrication workflows focuses on how to

F. Raspall (✉)
Graduate School of Design, Harvard University, Cambridge, USA
e-mail: craspall@gsd.harvard.edu

F. Amtsberg · S. Peters
Institute for Structural Design, Graz University of Technology, Graz, Austria
e-mail: felix.amtsberg@tugraz.at

S. Peters
e-mail: stefan.peters@tugraz.at

W. McGee and M. Ponce de Leon (eds.), *Robotic Fabrication in Architecture, Art and Design 2014*, DOI: 10.1007/978-3-319-04663-1_23,
© Springer International Publishing Switzerland 2014

develop a seamless, precise, and agile connection between digital models and production means, with the aim of achieving accurate physical products from digital models efficiently and effectively. Design-to-fabrication involves a one-way digital process from digital information to physical outputs. In this approach, two elements are crucial: a fluid exchange of information between geometric models and fabrication machines, and predictable material processes to secure expected results.

Through experimentation and prototyping, designers explore materials processes and build knowledge to anticipate how specific materials respond to the applied actions. In general, this knowledge enables a high degree of predictability.

However, the response of certain material processes, mainly due to their physical complexity, cannot be accurately predicted. This makes conventional design-to-fabrication workflows difficult to implement and limits the range of techniques that digital workflows can engage.

The addition of feedback control to design-to-fabrication workflows is achieved by introducing physical information before and during the fabrication process, using sensors. This information is employed to adjust the production process accordingly and accomplish the desired output. In this way, feedback control can expand CAD/CAM into more complex and irregular material processes.

2 Irregular Material Processes and the Need for Feedback in Robotic Production

Digital processes based on the strict formal language of computers are precise and predictable within their internal logic. In contrast, material processes supported by the physical complexity of reality always present a level of inaccuracy. Through industrialization and standardization of materials and processes, the margin of uncertainty is controlled through strict and well-defined industrial criteria, tolerances, and quality controls. An element that does not fulfill predefined standard is rejected. This provides designers with control over the materials with which they operate. CAD/CAM systems are very advantageous, increasing precision and efficiency.

The predictability of material processes differs according to the type of material and the transformation process that is exerted on it. For example, synthetic, industrial, and anisotropic EPS is a highly standardized and regular material. In addition, CNC milling of foam, in which a solid block is carved layer-by-layer through precise tools and motions, is very predictable (Fig. 1a).

In contrast, some materials are irregular and more sensitive to the environment, for example, wood or clay. Also, in certain material processes, tool actions not only affect the spot where the tool is acting, but also extend to surrounding areas or the whole assembly, making the process significantly more difficult to predict as inaccuracies accumulate in space and in time (Fig. 1b).

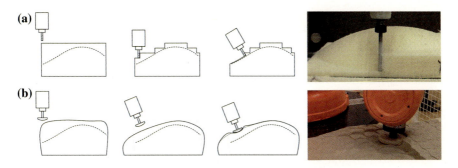

Fig. 1 CNC foam milling versus robotic clay forming. **a** In CNC foam milling, the material is highly controlled and inert, the tooling process is very precise, and the tool affects only the material engaged by the tool. **b** In robotic clay forming, actions affects not only the place where the tool is acting but the whole lump of clay

Fig. 2 Digital model of the formwork and physical prototype

These type of challenges have been encountered on several projects. The extensive work with ceramics at the Graduate School of Design at Harvard University (GSD) constantly faced the difficulties of working with clay, a malleable material that is easily deformed and very sensitive to environmental conditions. In addition, the plastic properties of clay make transformation processes, including wire-cutting, extruding and forming, hard-to-predict as tools acting on the material affect large areas of the block or assembly.

Another hard-to-predict process is part of the research at the Institute for Structural Design at the Graz University of Technology (TU Graz). The institute is developing a universal formwork, consisting on a rasterized field of pins covered with an elastic membrane. During the adjustment process, the exact geometry is very difficult to anticipate with enough precision, because the adjustment of each pin affect the whole surface (Fig. 2).

Comparable concerns with hard-to-predict material processes have been discussed in other research projects. Aggregate deposition, i.e. the pouring of thousands of particles, is being researched at the ICD Stuttgart by Dierichs et al. (2013). Piling of blocks of different heights is being investigated at the Federal Institute of Technology (ETH) in Zurich (Helm et al. 2012) in the framework of on-site robotics.

3 Overview of Mold-Making for Freeform Production

In order to test real-time feedback for robotic fabrication, a specific project serves as proof of concept. Complex-geometry designs, which are nowadays relatively simple to model, often encounter a definite limitation during the production phase: the production of thousands of one-off pieces is a time- and resource-consuming task.

A widespread solution is milling synthetic foams such as EPS, which are precise and easy to mill. Still, milling curved surfaces is time-consuming and generates substantial amount of waste: often more material is discarded than what is left in the piece. In addition, for high quality surfaces, materials can be expensive and require special coatings. The need for alternatives has been engaged in several projects for universal formworks.

The experiment presented in this chapter evaluates the possibilities of modeling clay to produce reusable molds. Plasticity, a key characteristic of this material, is the property a substance has when deformed continuously under a finite force. When the force is removed or reduced, the shape is maintained. This characteristic is, at first sight, promising for the creation of freeform molds. Rather than carving a mold from a solid block, malleable clay can be shaped into the target geometry. After the mold has been used, it can be reshaped.

Clay for artistic applications has been utilized extensively. In the industry, it is frequently used in modeling cars. Its application in robotic production has been introduced by Schwartz and Prasad (2013) for the production of molds for carbon-fiber products. Their subtractive method consists on carving a mass of clay with a wire-shape. While this proved to be a precise and efficient method for carving, it can be slow when freeform shapes have to be created from a large block of clay.

Significant research on ceramics has been developed at Harvard GSD, including robotic fabrication to extrude, cut, and form unfired clay and pick-and-place fired ceramic components. Some of the results were presented in Rob|Arch 2012 (Bechthold and King 2013).

Efforts to produce free-form molds for concrete are currently under development at ITE, TU Graz, including a large scale universal formwork for which the adjustment intends to implement feedback from a high-quality 3D scanner (Fig. 2). In the first semester of 2013, the master's design studio Faksimile, held at TU Graz, researched alternative mold making methods for freeform surfaces.

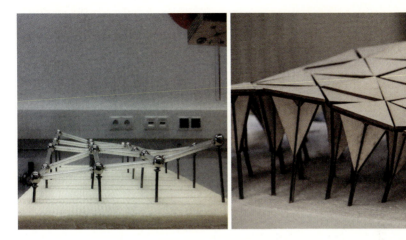

Fig. 3 Magnetic scaffold for moldmaking by Gerda Villgrater and Martin Bratkovics. Mold made out of triangular components by Pia Pöllauer and Lukas Jakober

Students explored a variety of mold-making strategies, including a telescopic 3D frame and a triangular-component mold system (Fig. 3). Within this studio, the project "Intricate Porosity", by Florian Landsteiner, proposed methods to form clay as reusable formwork for concrete casting. Two material processes were tested: forming a block of clay by pushing the clay with a "hammer" and carving the final shape using a shape-wire.

4 Overview of Material Feedback in Robotic Production

This research studies how feedback mechanisms can be implemented in current CAD/CAM processes, in ways that are open to and customizable by designers. The basic idea of feedback control is that the results of the latest action on the workpiece can be analyzed and used to determine the next action. Therefore, the conventional one-way process becomes an iterative sequence that includes analysis and adjustments during fabrication.

Feedback control in architectural robotics is being developed by some research groups. In addition to the already mentioned work by Helm et al. (2012), the research by Dörfler et al. (2013) studies feedback control in the framework of human–machine interaction.

Additive and subtractive material processes, the basic strategies in digital fabrication, are intimately related to feedback because aggregation and subtraction depend on what has previously done in order to build up or carve out the desired result. In an environment of highly standardized material processes, the whole procedure can be anticipated in advance and it is possible to execute the complete task without intermediate adjustments. However, when working with imprecise and hard-to-predict material processes eliminate, real-time material feedback becomes a necessity. In order to introduce feedback control to current design to fabrication workflow, four additions or modifications are required:

4.1 Sensors: Feedback Signal

First, feedback control requires sensors to gather information from the physical world. Industrial robots have built-in sensors that e.g. measure and adjust the drive current but for some industrial applications external sensors are used, such as contact sensors and vision systems for mounting wheels on cars or picking and placing food. Programming, calibration and control of external sensors is an advanced task and the implementation in architectural robotics has been limited.

4.2 Control Protocols

Second, the control system requires algorithms that determine how instructions are adjusted in response to the information gathered by sensors. Rather than a fixed and complete set of instructions, fabrication is defined by decision rules, algorithms, which determine adjustments in real-time eliminate. The process is divided into iterations: At each iteration, relevant aspects of the assembly are measured and used to adjust the instructions. The process runs until the assembly matches the desired target.

4.3 On-Line Machine Operation

Third, at each cycle, new calibrated instructions are sent to the robot. Regular off-line operation of robots, which requires manual upload and execution of instructions, becomes too inefficient. On-line operation of the robot is required.

4.4 Compatible Information Platform

Finally, a compatible platform across sensing, actuating and controlling devices is necessary to ensure a consistent flow of information.

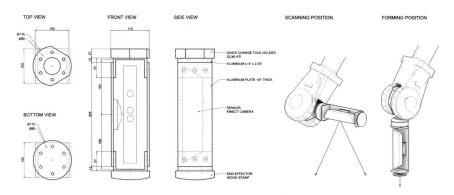

Fig. 4 Tool and camera detail. Example of typical use

5 Material Feedback for Robotic Production of Freeform Reusable Molds

"Intricate Porosity" showed initial feasibility for clay forming processes. However, the implementation of feedback control significantly improves speed and accuracy. The forming process without feedback required very small increments to keep the displacement of the clay under control.

The project starts from a similar material process, adding a feedback control mechanism to the workflow. The following sections describes the components of the project.

5.1 Robotic Arm, Tool and Sensor

The feedback project employs an ABB1440 robot with a custom tool which serves both as sensing and actuating device (Fig. 4). The material used is white stoneware.

5.2 Control Algorithm, Information Exchange Platform and On-Line Control Protocol

The clay forming process implements a feedback-controlled sequence of iterations, each consisting on three steps. Figure 5 illustrate the steps, described here:

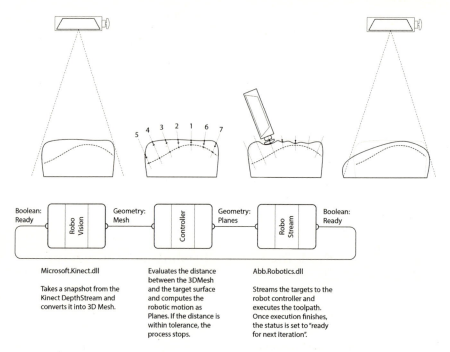

Fig. 5 Diagram of the clay forming process

As first step, the 3D camera provides a point cloud representation of the existing clay block. The initial lump of clay does not need to be precise, only slightly bigger than the desired geometry. The camera information is analyzed and only the points that belong to the block are passed to the control algorithm. The EMGU wrapper for OpenCV computer vision was implemented to perform this simple recognition task.

The second step involves the evaluation of the difference between the existing material and the desired surface. The algorithm divides the surface into areas slightly smaller than the hammer tool, and calculates the trajectory line from the existing block of clay to the desired target surface. Then, it determines the pushing motions considering the depth that the tool penetrates into the clay. The order in which the pushing motions are executed is determined by weighting three different criteria: from the areas where clay is in excess to the areas where clay is missing, from the center to edges, from hills to valleys.

In a third step, the instructions are sent to the robot controller and executed. In order to stream the instruction from the modeling platform to the robot, an implementation of the procedure proposed by Cederberg et al. (2002) was programmed.

After the motions have been completed, the process is iterated, but now the clay block is closer to the target geometry. The process stops when the target surface and the real surface match within an acceptable predefined tolerance. If sectors of the real surface lie below the target when the rest of the clay matches the target, the system request the manual addition of clay.

The whole process was programmed in C# as custom McNeel's Grasshopper components.

5.3 Applications for Clay Molds

The clay molds developed in this experiment are used for supporting fabric concrete and for casting concrete. In addition, clay molds can be used for fiber reinforced polymers and as reusable scaffold for freeform assemblies.

6 Conclusions and Outlook

In the primary experiment, forming clay, a material with malleable properties but hard to operate using the one-way design-to-fabrication paradigm, was successfully implemented to produce freeform molds. Using a similar workflow, feedback control can be extended to a wide variety of material processes—e.g. wood and steel bending where spring back is hard to anticipate, universal formwork adjustment, among others.

Several challenges were faced during the development of the experiment. First, programming the 3D-camera interface and online control of the robot for Grasshopper was a laborious task. In a similar way as HAL, Rhino2krl, and Your (Schwartz 2013; Pawlofsky 2013; Lim et al. 2013) expanded robotics to a sizeable number of designers, similar efforts to systematize and simplify programming of feedback control will surely open this approach to larger audiences.

The vision system in this project is a relatively simple procedure to detect which points belong to the clay block and which are background. These 3D points are directly used by the control algorithm. The computer vision libraries used in this research can perform more advanced tasks such as curve reconstruction, which might be needed in other applications.

Feedback control replaces fixed instructions with decision rules, making the process responsive to material information. In this way, certain degree of autonomy is transferred to the system, limiting the capacity of designers to simulate or control the process unequivocally. Therefore, the duration and achievability of a

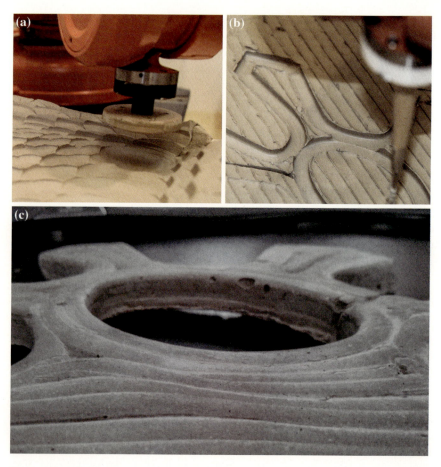

Fig. 6 a Robotic forming of clay block. **b** Robotic carving of the negative. **c** Detail of a final piece in concrete

task cannot be fully predicted. In addition, feedback systems are susceptible to several issues such as oscillation traps, in which the process gets stuck between opposed adjustment responses. Careful testing and calibration of the processes are required.

The use of feedback systems in computer aided fabrication for architecture can open the range of possible materials and fabrication methods to the designers. In a relatively simple case as clay forming, it eliminate proved successful (Fig. 6). It is the intention of the researchers to systematize the tools and processes and deploy them in more complex material processes.

Acknowledgments The research is supported by the Design Robotic Group at GSD Harvard and the ITE at TU Graz. Assistance from Panagiotis Michalatos, from Harvard GSD, was essential in the programming of the different components of the process.

Projects from the master studio Faksimile presented in the chapter are the result of the work by Gerda Villagrater, Martin Bratkovics, Pia Pöllauer, Lukas Jakober and Florian Landsteiner.

Programming during this research involves a variety of free and academic-licensed software, including Grasshopper, Kinect SDL, EMGU and OpenCV. In early experiments, Thibault Schwartz' HAL (2013) was instrumental to create the robot motions.

References

Bechthold M, King N (2013) Design robotics. In: Brell-Cokcan S, Braumann J (eds) Robotic fabrication in architecture, art and design. Springer, Wien, New York, pp 118–130

Cederberg P, Olsson M, Bolmsjö G (2002) Remote control of a standard ABB robot system in real time using the robot application protocol (RAP). In: Proceedings of the 33rd international symposium on robotics (ISR), 7–11 Oct 2002, paper no. 113

Dierichs K, Schwinn T, Menges A (2013) Robotic pouring of aggregate structures: responsive motion planning strategies for online robot control of granular pouring processes with synthetic macro-scale particles. In: Brell-Cokcan S, Braumann J (eds) Robotic fabrication in architecture, art and design. Springer, Wien, New York, pp 196–205

Dörfler K, Rist F, Rust R (2013) Interlacing: An experimental approach to integrating digital and physical design methods. In: Rob-Arch 2012: Robotic Fabrication in Architecture, Art and Design

Helm V, Selen E, Fabio G, Matthias K (2012) In-situ robotic construction: Extending the digital fabrication chain in architecture. In: Proceedings of the 32nd annual conference of the association for computer aided design in architecture (ACADIA), ACADIA, San Francisco, pp 169–176

Lim J, Gramazio F, Kohler M (2013) A software environment for designing through robotic fabrication. In: Proceedings of conference on computer-aided architectural design research in Asia (CAADRIA 2013), pp 45–54

Pawlofsky T (2013) Rhino2krl. In: Brell-Cokcan S, Braumann J (eds) Robotic fabrication in architecture, art and design. Springer, Wien, New York, pp 167–172

Schwartz T (2013) HAL: extension of a visual programming language to support teaching and research on robotics applied to construction. In: Brell-Cokcan S, Braumann J (eds) Robotic fabrication in architecture, art and design. Springer, Wien, New York, pp 92–101

Schwartz M, Prasad J (2013) Robosculpt. In: Brell-Cokcan S, Braumann J (eds) Robotic fabrication in architecture, art and design. Springer, Wien, New York, pp 230–237

Phase Change: Approaching Research Methodologies Through Design Robotics

Opportunities for Customization Through Robotic Manipulation of Materials During Phase Change and Curing Processes

Nathan King, Kadri Tamre, Georg Grasser and Allison Weiler

Abstract This chapter positions ongoing research surrounding robotic manipulation of materials during phase-change and curing conducted by the University or Innsbruck Institute for Experimental Architecture's REX|LAB within the context of Design Robotics, a research paradigm established by the Design Robotics Group at the Harvard University Graduate School of Design. The goal of this analysis is the identification of key research principles and evaluation criteria that can inform future developments toward the low-volume production of polymer-based architectural building components. An overview of industrial plastic-forming techniques is used to identify limitations and opportunities within the associated material systems and the findings used to contextualize examples from ongoing research. A series of strategic next steps are proposed and will be explored during a series of workshop experiments.

Keywords Design robotics · Material feedback · Material processes · Synchronized robotic fabrication · Design research

N. King (✉)
Harvard GSD Design Robotics Group, Cambridge, MA, USA
e-mail: NatlKing@VT.edu

Virginia Tech CDR|CRAFT, Blacksburg, VA, USA

Rhode Island School of Design, Providence, RI, USA

K. Tamre · G. Grasser · A. Weiler
Institute for Experimental Architecture REX|LAB, University of Innsbruck, Innsbruck, Austria
e-mail: Kadri.Tamre@uibk.ac.at

G. Grasser
e-mail: Georg.Grasser@uibk.ac.at

A. Weiler
e-mail: Allison.Weiler@uibk.ac.at

W. McGee and M. Ponce de Leon (eds.), *Robotic Fabrication in Architecture,*
Art and Design 2014, DOI: 10.1007/978-3-319-04663-1_24,
© Springer International Publishing Switzerland 2014

1 Introduction

Ongoing research at the University of Innsbruck Institute for Experimental Architecture's REX|LAB has begun to realize the provocative potential of material manipulation during phase change or curing through design experimentation and course work relating to the continuous synchronized robotic material manipulation (Colletti 2013). These investigations exploit perceived limitations within common polymer-forming processes that typically shape material using a fixed-form mold. In part a result of tooling costs relative to production volume, the architectural application of plastics is largely relegated to high volume sheet-goods, coatings, and fiber-based composites. Where mass produced mold-formed parts are used, the potential for customization is extremely limited. The design research begun in the REX|LAB attempts to establish a new paradigm for polymer forming through the use of synchronized industrial robotic material manipulation that utilizes real-time feedback to shape materials during phase change or curing (Tamre et al. 2014).

Design Robotics, as a mode of inquiry established by the Harvard University Graduate School of Design, Design Robotics Group, suggests a design research methodology that attempts to isolate strategic research opportunities within a given material system through rigorous evaluation of materials, processes, and industry protocols using strategic design experiments. "This…form of research seeks to bridge the gap between primarily artistic endeavors and an often-conservative industry that has long occupied the tail end of manufacturing innovation" (Bechthold 2013). In the context of strategic process intervention this effort seeks to establish a trajectory that will help to inform the future developments of the ongoing REX|LAB investigation into temporal forming of polymeric materials.

2 Polymer-Based Materials Systems

When developing processes for architectural applications, as with all product design, parameters such as part size, production volume, tooling costs, assembly logic, and material properties must be considered. Through the analysis of polymer-based material systems, opportunities for architectural application and customization were identified.

2.1 Phase Change and Curing

In the context of this research polymeric materials or plastics refers broadly to many polymer-based materials, both thermoset and thermoplastic, formed using a variety of processes. Thermoplastic materials are available in a variety of formats from chips to sheets and can be formed and re-formed under heat. Thermosets, however undergo a permanent chemical change process during curing that cannot be reversed. In the context of this research the term '*Phase-Change*' applies to thermoforming materials that change state during heating and or cooling. Curing is used to define the chemical change that occurs when a material transitions from a liquid resin to a solid though chemical reaction.

2.2 Systemic Limitations

Through survey of polymer-based materials it was determined that limitations in tooling must be resolved to expand the potential realization of low-volume architectural applications. In all surveyed manufacturing processes phase change or curing is utilized in the delivery of the material to the tool and where the material is held static to the mold surface. In most additive processes, once the material is placed or cured, no additional material manipulation is realized until post-production. Systemic limitations in tooling present two immediately viable research trajectories; a problem based approach that addresses the demand for low volume production of polymeric building components and a range of opportunity-based speculation toward the manipulation of materials during phase-change.

3 Ideation: REX|LAB Experiments

3.1 Prototypical Experimentation and Ideation

The following experiments conducted by researchers at the REX|LAB begin to address tooling limitations through a variety of robotic forming processes that limit the need for specialized molds. Prior to automation a series of manual operations began to impart an understanding of material behavior in response to stimuli including, temperature, force, etc. Automated material experiments were conducted using an industrial robotic work cell consisting of three synchronized robotic manipulators and a variety of project specific end effectors. Still in early stages, these experiments address primary forming as a variable process by manipulating the material during a phase change. A selection of strategic process experiments that have been conducted:

Robotic Foaming is an experimental fabrication process that takes advantage of material expansion during curing of polyurethane foams. Here, an automated robotic process collects material and during the curing process, stretches it between multiple robotically controlled surfaces. Resulting forms are organic in appearance and resemble the results of a physical topology optimization process

Fig. 1 Robotic foaming process exploiting for maximum dimension, conducted at Smart Geometry 2013. *Images* REX|LAB

suggesting that there may be structural implications of the material behavior. Environmental conditions among other parameters make exact results difficult to predict but average cross sectional dimensions suggest that the process is repeatable within a given tolerance (http://smartgeometry.org/index.php?option= com_community&view=groups&groupid=38&task=viewgroup). Additional feedback strategies were tested that can be used to collect process data during fabrication and adjust manufacturing parameters in real-time (http://responsiverobotics. wordpress.com/) (Fig. 1).

In response to tooling limitations ***Robotic Thermoforming*** experiments were conducted using three collaborative industrial robotic manipulators to locally deform thermoplastic sheet materials using custom end effectors to apply heat, position the stock, and deform the surface. This process was tested through physical prototypes. Similar to other sheet forming processes limitations were encountered in relation to maximum deformation and associated material thickness (http://vimeo.com/61215586) (Fig. 2).

Thermoplastic Stretching, similar to Robotic foaming, utilizes multiple end effectors that collect a molten thermoplastic billet and stretch it as it cools. Current tests use polypropylene, but other plastics are possible. In contrast to robotic foaming, no material expansion is seen and the resulting parts exhibit increased compression resistance consistent with known material properties (http://www. exparch.at/wiki/index.php/POFC_-_Process_of_forming_Chitin) (Fig. 3).

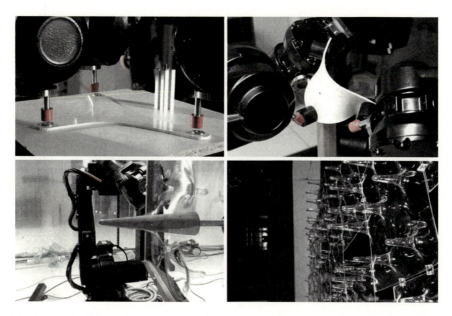

Fig. 2 Localized thermoforming experiments conducted by the REX|LAB in 2012. Here a more precise relationship between designed surface and the process outcome is possible. *Images* REX|LAB

Fig. 3 "Thermoplastic stretching" material study utilizing polypropylene conducted during the course 'Meeting Nature Halfway'. *Images* REX|LAB

4 From Ideas to Concepts: Evaluation and Development

4.1 Evaluation of Material Experiments

While these ideation exercises are at various stages of development and precise conclusions cannot be made without further testing and evaluation, they can be used to inform a series of next steps. Typical evaluation criteria for Design Robotics research include the following (Bechtold and King 2012):

- What kind of customization can be achieved with the process? All processes evaluated enable a high degree of customization within a given set of constraints.
- What are the mutual dependencies between customized parts and the overall installed system? At this stage of development most processes cannot yet be viewed from a systems perspective but in the more developed processes, robotic foaming for example, the end condition or final target position is the most obvious mode of engagement within a larger system. For example, the process has been used to develop nodes within a spatial structure. In current examples of localized thermoforming the edge condition is predetermined enabling system integration at a global scale.
- What value does part customization present for the material system and potential applications? All processes have been developed based on recognized material and process opportunity surrounding automated digital fabrication processes. Limitations in the existing material system support the notion that robotic forming could enable low volume production of architectural building parts.
- Can waste be reduced? Because all tested processes form material without molds it is likely that a reduction of waste is possible when compared to alternative customization processes where individual molds are created for each part. Furthermore it is conceivable, but yet to be evaluated, that some of the strategies may enable performance-based material optimization. It is important to note that at the moment they are merely visual observations.
- How are installation procedures affected by part variation? As with most customization processes, assembly and installation logic must be developed. In the installed example of robotic foaming differentiated parts were labeled for identification but limitations in relational orientation present unique installation challenges. A key design challenge in this system is to embed assembly logic into the tooling. Other processes maintain uniform edge conditions so location data is enough to communicate installation sequencing.
- To what degree can the process be incorporated into state-of the art industrial production lines? Early process experiments present new forming paradigms that may not fit within existing production lines. These processes seem to present new manufacturing opportunities that may be incorporated where a material supply is available, thus potentially incorporated in existing

manufacturing environments. It is likely that these processes would emerge within more agile production environments where the robot can be used for a range of applications. In the case of localized material deformation, industrial integration is possible in most sheet-based thermoforming facilities.

4.2 From Evaluation to Evolution

Comparison to the state of the art in material systems can help direct further research but may yield only part of the answers. Ideation and speculation leads to more questions that help inform specific research areas and trajectories. Based on the above evaluation it has been determined that application potentials should be evaluated and a next step would be to identify a series of design and redesign experiments that test application potential for a given process. Through this study the group will gain a better understanding of the process opportunities and limitations. Within this research the following questions should be addressed and used to reevaluate the overall research trajectory.

- What are the primary processes and material properties that need to be understood to begin to establish application potential?
- At what point is the process being developed a simulation of some other industrial condition, material, or process and at what stage is the robotic process a proposal for a completely novel mode of material manipulation?

5 Conclusions

Early explorations and ideation can be categorized in the context of the related material systems to inform basic research questions leading to eventual application of novel processes; either as new industrial opportunities or as part of strategic process intervention within a given manufacturing scenario. Here a specific limitation, tooling, is an obvious hurdle to widespread use of various plastics and specifically application of individualized forms. The search for pragmatic solutions can often stifle exploration therefore suspended judgement is necessary to insure the explorative quality of early phases of testing. With this in mind specific questions have been identified above that can be used to inform next steps in the ongoing research. These questions will be the focus of a series of workshops that explore phase change and curing in the context of design robotics.

Acknowledgments Research support provided by Marjan Colletti, chair of the Institute for Experimental Architecture; Development of synchronous robotic work cell supported in part by ABB-Austria; Research consultation provided by Harvard GSD Design Robotics Group and the Virginia Tech Center for Design Research-CRAFT Group; Automated synchronous robotic programming workflow, HAL, developed by Thibault Schwartz.

References

Bechthold M, King N (2012) Design robotics: towards strategic design experiments. In: Brell-Cokcan S, Braumann J (eds) RoblArch 2012—robotic fabrication in architecture, art and design. Springer-Verlag, Vienna, pp 118–129

Bechthold M (2013) Design robotics: new strategies for material system research. In: Peters B, Peters T (eds) Inside smart geometry. Wiley, London, pp 254–267

Colletti M (2013) Protorobotic foaming. Archithese 3:54–57

Tamre K, Colletti M, Grasser G, Weiler A (2014) (Fr)Agile Materiality, (to be published in Fabricate 2014 Conference), 14 Feb 2014

Sense-It

Robotic Sensing and Materially-Directed Generative Fabrication

Ellie Abrons, Adam Fure, Alexandre Dubor, Gabriel Bello Diaz, Guillem Camprodon and Andrew Wolking

Abstract *Sense-It* explores the potentials of *materially-directed generative fabrication* through an integration of research in robotic sensing, plastic deposition, and generative code. This approach tests the limits of a machine-material-sensor interface to act autonomously, without direct adjustments from an observing operator, and capitalizes on sensor responsiveness and material agency to produce unpredictable outcomes. The research moves away from optimization and efficiency as the primary drivers of digital fabrication in pursuit of a model where materials assume maximum agency in the fabrication process. Feedback loops between machining parameters, real-time sensors, and plastic deposition infuses the work with both intelligence and an intentional instability, where the outcomes can be guided but never fully predicted.

Keywords Robotic fabrication · Real-time sensing · Material agency · Plastic deposition

E. Abrons (✉) · A. Fure · A. Wolking
University of Michigan, Ann Arbor, MI, USA
e-mail: eabrons@umich.edu

A. Fure
e-mail: afure@umich.edu

A. Wolking
e-mail: awolking@umich.edu

A. Dubor · G. Camprodon
IAAC/Fab Lab Barcelona, Barcelona, Spain
e-mail: alex@fablabbcn.org

G. Camprodon
e-mail: guillem@fablabbcn.org

G. B. Diaz
Future Architectural Coalition, Seattle, WA, USA
e-mail: gabriel@makeitlocally.com

W. McGee and M. Ponce de Leon (eds.), *Robotic Fabrication in Architecture,*
Art and Design 2014, DOI: 10.1007/978-3-319-04663-1_25,
© Springer International Publishing Switzerland 2014

1 Introduction

As architectural research into robotic fabrication advances, various models of interaction between designers, digital-modeling software, machine, and material are emerging. As technical expertise builds and spreads throughout the architectural community, designers are not only experimenting with new tools and processes, but with entirely new conceptions of the role of the designer within such processes. In other words, the community advancing architectural robotics is quickly moving away from a consensus on the 'right' way to approach research to the experimentation and articulation of many 'possible' approaches. Each of these approaches carries with them both specific techniques and conceptual attitudes. In this chapter we will outline the approach and attitude of the *Sense It* workshop to be held at RobArch2014.

Sense It explores the potentials of *materially-directed generative fabrication*. This approach gives materials agency in the design process whereby the physical attributes of unpredictable materials are sensed in real-time in order to direct subsequent tool paths. This approach to fabrication is distinct from those that rely on very small tolerances of material behavior in order to carry out a lengthy set of predetermined tool paths. Our approach cedes design control to a closed loop of robotic tool paths, unpredictable material behavior, and real-time sensing technology. Tool paths are not restricted to a predetermined set of movements or human-controlled real-time adjustments; instead, real-time sensing data is used to adjust tool paths during the fabrication process with no immediate intervention from a human agent. Such an approach tests the limits of a machine-material-sensor interface to act autonomously, without direct adjustments from an observing operator and capitalizes on sensor responsiveness and material agency to produce unpredictable outcomes.

2 Key Concepts

2.1 Anti-Optimization

Materially-directed generative fabrication necessitates a shift away from an epistemological model of optimization, the dominant paradigm within digital fabrication. In short, research governed solely by optimization checks progress against numerous efficiencies (e.g., structural, thermal, or acoustic). Results of the fabrication process that cannot be justified in terms of improving a measure of efficiency must be eliminated. Removing the imperative of optimization allows for unexpected results of the fabrication process, such as new types of figuration derived from material behavior as in the case of "magnetic architecture."(Dubor and Gabriel-Bello 2012) (Fig. 1), to breed new focuses for research. These focuses may break from the principles of optimization: they may be structurally inefficient

Fig. 1 Robotically-generated metal forms by Gabriel Bello-Diaz, Alexandre Dubor, Akhil Kapadia and Angel Lara

or materially excessive, necessitating new value sets and modes of analysis. Allowing other epistemological models to emerge through experiments in robotic fabrication assures a future where the possible applications of these technologies are constantly expanding outward as opposed to converging on a single, perfected model.

2.2 Unpredictable Material Agency

"On the one hand, to the formed or formable matter we must add an entire energetic materiality in movement, carrying singularities or haecceities that are already like implicit forms that are topological, rather than geometrical, and that combine with processes of deformation: for example, the variable undulations and torsions of the fibers guiding the operation of splitting wood. On the other hand, to the essential properties of the matter deriving from the formal essence we must add variable intensive affects, now resulting from the operation, now on the contrary making it possible: for example, wood that is more or less porous, more or less elastic and resistant. At any rate, it is a question of surrendering to the wood, then following where it leads by connecting operations to a materiality, instead of imposing a form upon a matter (Delueze and Guatarri 1987)"

The importance of materials within digital production has increased with the shift from the exclusively software-based formal experiments of the early years, which took place in the immaterial, virtual platforms of design software, to the fabrication of physical prototypes, which by necessity, account for real material constraints. Early research into digital fabrication relied on commercially-produced stock materials whose tolerances could be accurately predicted, thereby minimizing unexpected material behaviors within the fabrication process. This early research closely resembles software-based digital design in which material is inert, formed by geometrical instructions imposed by the user (Oxman 2012). More recent research, such as that conducted by Menges (2012) at the University

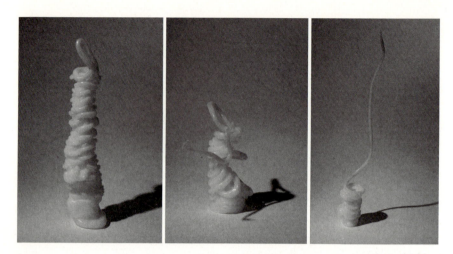

Fig. 2 Variations of plastic forms based on material behavior (*Images* Justin Tingue and Andrew Wolking)

Of Stuttgart Institute Of Computational Design, is beginning to integrate material behavior into fabrication processes, and initiating a new focus on material agency within digital fabrication. Such research has been theoretically inspired by a persistent aspiration of computational design, long theorized by French philosopher Gilles Delueze and extended in the past decade by Manuel De Landa: to collaborate with matter to produce form (De Landa 2001). Crucial to this aspiration is the recognition of material agency, that is, the ability of material to produce or inflect form all on its own, with little to no human intervention.

Agency, or the ability to intercede in a process, is a relative term, and can refer to a broad spectrum of influences with wide-ranging degrees of force. The aforementioned research of Menges and his colleagues represents a small portion of that spectrum. In order for their research to be successful, material behavior must be numerically quantifiable so that it can be input into digital simulations that precede fabrication. The end result is a combination of material behavior and computational form, but within a very small scope of possible materials and behaviors. *Materially-directed generative fabrication* explores the potential of unpredictable material behavior to generate form. "Unpredictable," in this case, means very large tolerances of material behavior, such as the wide range of vertical forms produced by plastic deposition (Fig. 2) This approach is related to "Digital Materiallurgy" as proposed by Fure (2011), which "relies on intentionally ceding limited design control to unpredictable matter—thus capitalizing on matter's innate ability to produce unexpected formal and material complexity".

2.3 Sensor Responsiveness

High tolerances of material behavior make long sequences of pre-determined tool paths difficult to carry out. Sequential tooling necessitates predictable material outcomes. For example, in RPD (Robotic Plastic Deposition, a research project explained in depth below), to initiate a plastic deposit on top of another deposit requires accurate prediction of the preceding deposit's location, which is difficult given the unpredictable nature of the plastic. The introduction of sensing technology, however, provides real-time data on material behavior, enabling the workflow to adjust with changes in the material output. Thus, sensors replace predetermined sequential tool paths or real-time adjustments from a human agent (Johns 2012).

The unique capacities of sensing technology enable material behavior to be generative. Typically, generative design in architecture refers to computer algorithms that generate form. *Materially-directed generative fabrication* enables material behavior to generate form. In other words, our process makes materials, and their constituent behaviors, the principal determinant of form, more influential than geometric constraints, numeric inputs, or computer algorithms. In this context, the research of Menges et al. might be considered "materially-influenced" fabrication, while our research is "materially-generative" fabrication.

3 Set-up

The *Sense It* workshop to be held at RobArch2014 combines robotic plastic deposition with temperature and distance sensing as a first test case of *materially-directed generative fabrication*.

3.1 Robotic Plastic Deposition

Robotic plastic deposition (RPD) is a digital fabrication process that uses a specialized end effector on a 6-axis robotic arm to melt and extrude polypropylene pellets. Initial research into RPD has focused on vertical structures and viscous figuration, two categories opened up by plastic's capacity to rapidly harden from a molten state into unpredictable formations (Fig. 3). This distinguishes it from other deposition materials such as clay, concrete, or sodium acetate that either require formwork or produce constrained and predictable forms such as piles or tubes. Distancing the end of the extruder from the base allows for the plastic to take on a variety of forms, producing eccentricities that act as catalysts for *materially-directed generative fabrication*.

In physical terms, the specialized end effector melts polypropylene pellets into a viscous stream that is extruded through a custom milled aluminum nozzle. The

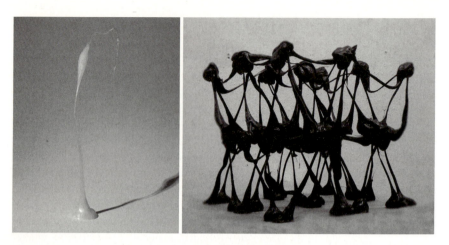

Fig. 3 Prototypical RPD vertical member (*left*) and space-frame test (*right*)

shape and size of the nozzle opening influence the formation of the plastic bead, or string that is deposited by the robot. This bead hardens and becomes rigid within seconds of deposition, producing physical snapshots of pouring liquid. Alternatively, pausing the robot's retraction allows the viscous plastic to build up in place, producing a glob of material that can act as a structural node or anchor for additional deposits. If the hardened forms come into contact with newly deposited plastic it partially melts, allowing formations to build up vertically in sequence and subtly influence final formations over time.

Preliminary research uses vertical lines as first-level test mechanisms. A series of initial deposits test the performance of the extruded plastic under changing variables such as geometry input and feed velocity (the speed of movement of the robotic arm). These studies are able to produce vertical formations through depositions lasting three seconds. Canceling the extrusion process in the code after an initial deposition while retracting the robot pulls the plastic vertically, due to the plastic adhering slightly to the aluminum extruder tip. The resultant height of each plastic column varies for each test despite consistent tool path heights, thus exhibiting the range of material behavior. Going forward, using sensors to detect the height of these plastic extrusions will allow adjustment of the tool path in real time, giving agency to the material in the fabrication process.

To technically understand the process of RPD it is important to understand the relationship between the geometric input, the script, and the robotic arm. First, a vertical line is created in Rhino3D model space. Using Supermatter Tools (SMT), developed by Wes Mcgee, Dave Pigram et al., the programmer is able to create a cohesive environment between machinic output and digital space. SMT provides a robotic simulation environment in Rhino to aid the user in predicting the tool paths of the robotic arm.

A series of prompts in the SMT interface aids users in programming the tool paths. To begin, the user calls the extruder and saves the output Gcode. Next, the

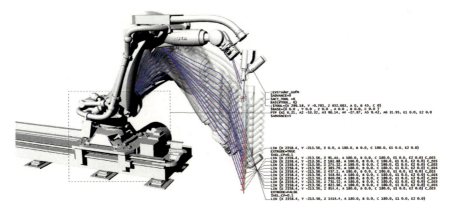

Fig. 4 Output diagram using supermatter tools (SMT) illustrating output machine code from Python Script of RPD of one 3 foot vertical line with 10 division points

base values of the robot are entered by reading its current location from the Kuka remote in real time, or by inputting its current location in Rhino3D's world coordinates. These P2P values give coordinates to the digital model and can be translated in real time. Once the location is set, the input curve geometry is selected where the user has the option of how many points to divide the curve into (Fig. 4). For vertical lines a low resolution line can be used with minimal divisions.

SMT next prompts the user to select the orientation of the robot from a number of options. "Fixed Orientation" keeps the extruder vertical as it traverses over a curve. Entering the feed rate gives the programmer control of the speed that the robotic arm moves over the geometry. Moving the arm too fast will stretch the melted plastic too thin and break the vertical formation. A feed velocity of .005 m/s allows the plastic to dry in midair and be stretched vertically. Once these steps are complete, the user has the option to pose the robot and visually verify in Rhino model space the path of the robot. This ensures that the input geometry is located within the limits of the robot. Figure 5 illustrates a range of tests to better understand the influence of process variables on the resultant material behavior.

3.2 Robotic Sensing

The setup for initial experimentation in *materially-directed generative fabrication* uses two synchronized 6 axis robots (Kuka KR 6 AGILUS) combined with a fixed plastic extruder and two sensors. An infrared thermometer (MLX90614) provides real time feedback on the temperature of the plastic indicating when it is likely to break from the extruder bead, while an infrared short range proximity sensor (Sharp GP2D120XJ00F) detects the height of the extrusion. The sensor values are transmitted via an Arduino microprocessor and then sent to the Kuka Robot

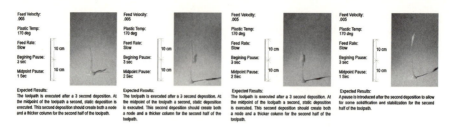

Fig. 5 Initial tests of RPD vertical extrusions

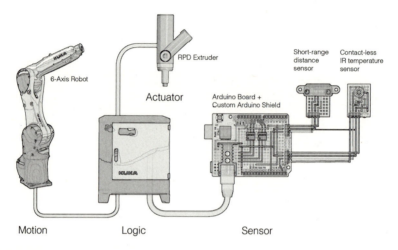

Fig. 6 Setup diagram/flow of information

Controller input/output bus through an optically isolated circuit. This integrates the sensors as regular Kuka signals, accessible as inputs on the KRL (Kuka Robot Language), scripts (Fig. 6).

3.3 Scripts

The logic of the generative process is encoded within the Kuka Robot Controller in KRL (Kuka Robot Langauage) using simple conditional loops and Function Generators (a technology developed by Kuka). Sensor feedback is already used in industrial applications, such as welding, in order to correct the trajectory or velocity of the robot to match the imprecision of a process with the expected precision of the robotic fabrication. *Sense It* utilizes the same robust technology to explore *materially-directed generative fabrication*. Additional scripts developed in previous research will be used to simplify and secure the setup (e.g., including a speed and a bounding box limit as well as basic routines for robot and extruder initialization). Using Arduino as a sensor middleware allows us to maintain simplicity within the KRL script, keeping the focus on the generative logic derived from the material.

4 Application

4.1 Logic 1

The aforementioned plastic extrusion tests, done without robotic sensing, highlight the large variation in the maximum height of a vertical extrusion, despite consistent tool paths. This exhibits the range of the material's behavior. The first fabrication logic of *Sense It* focuses on a vertical movement using the temperature sensor to control the extruder (heater on/off, feed rate) and to control the speed of the robot (from 0 to 0.05 m/s). The temperature sensor will determine in real-time how much the material can be stretched and eventually detect a lack of material (temperature threshold). The outcome of this experiment will be a set of vertical formations with variable density, height, and identifiable patterns, expressing the generative logic of the code (Fig. 7).

4.2 Logic 2

Logic 2 involves connecting multiple extrusions. This setup uses a distance sensor to guide the extruder in depositing material layer-by-layer, accumulating material in the Z-axis. In the XY axes, a concentric series of points is pre-encoded and repetitively used as starting point for each extrusion. The extruder then moves in the direction of the center of the base, checking at each step the height of the existing depositions for the next position. This incremental logic allows the deposition to be right on top of the previous extrusion and insures a strong connection. This process is executed until a specific height is reached. Here the initial points and the specified height target define the global underlying geometry while the material behavior determines the idiosyncrasies of the resultant formation (Fig. 8).

5 Conclusion

Materially-directed generative fabrication multiplies the possible uses of established technologies (plastic deposition, industrial robotic sensing, and generative coding) to explore the limits of machine-material-sensor interfaces. The ability to sense material behavior in real-time radically expands the potentials of matter to

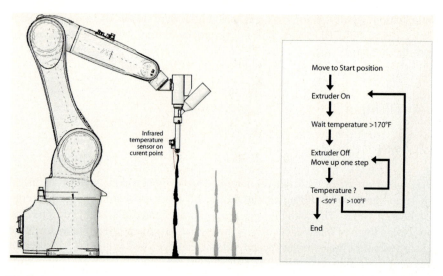

Fig. 7 Logic 1 diagram

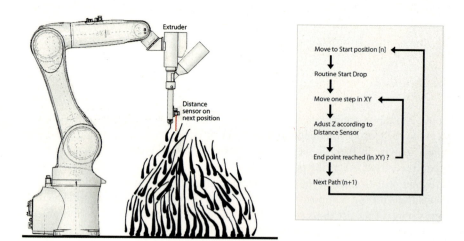

Fig. 8 Logic 2 diagram

inform fabrication and influence form. Sensors allow materials to assume maximum agency in the fabrication process, no longer ignored or utilized only in limited capacity. While the specific applications of this research in the context of the Sense-It workshop focus on temperature and distance sensing combined with plastic deposition, it is our intention that *materially-directed generative fabrication* enters the robotic fabrication community as a paradigmatic model of research, one that is open to other machines, materials, sensors, and ultimately, other architectures.

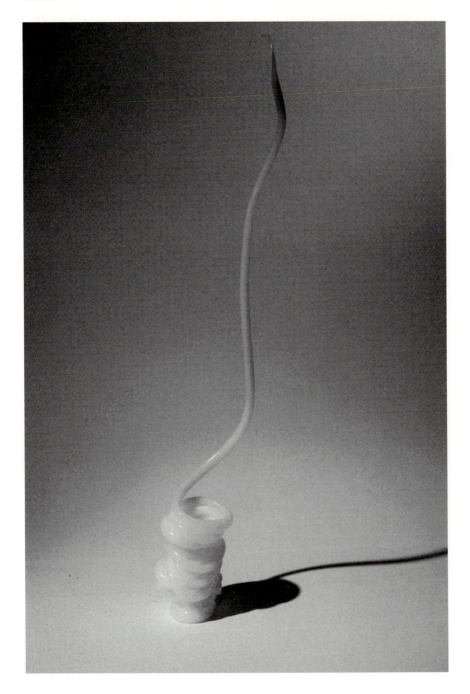

References

De Landa M (2001) Philosophies of design: the case of modelling software. In: Salazar J (ed) Verb processing. Actar, Barcelona, p 132

Deleuze G, Guattari F (1987) A Thousand plateaus (trans: Massumi B). University of Minnesota Press, Minneapolis, p 408

Dubor A, Gabriel-Bello D (2012) Magnetic architecture: generative design through sensoric robots. In: Rob|Arch Robotic Fabrication in Architecture, Art, and Design, Springer-Verlag, Vienna, pp 206–213

Fure A (2011) Digital Materiallurgy: on the productive force of deep codes and vital matter, ACADIA 11: Integration through computation. In: Proceedings of the 31st annual conference of the association for computer aided design in architecture (ACADIA), Banff (Alberta), 13–16 October 2011, pp 90–97

Johns R L (2012) Augmented reality and the fabrication of gestural form. In: Rob|Arch robotic fabrication in architecture, art, and design, Springer-Verlag, Vienna, pp 248–255

Menges A (2012) Material Resourcefulness: Activating Material Information in Computational Design. Architectural Des 216:34–43

Oxman N (2012) Programming matter. Architectural Des 216:88–95

Part IV
Industry Papers

KUKA Robots On-Site

Stuart Shepherd and Alois Buchstab

Abstract Within a relatively short time span, industrial robots have turned from an exclusive tool for high-end industries to a multifunctional machine for a wide range of users, from small and medium enterprises to architectural and industrial design offices. The demands of these users differ significantly of how robots are used in the automotive industry, with robots working on-site and even in the close proximity of human workers. New technological advances are required to cover these needs, from light-weight robots with an array of sensors to mobile platforms that can autonomously move heavy robotic arms with millimeter accuracy.

Keywords KUKA · Lightweight robot · Soft robotics · Mobile robot platform · Robotic arms · Man–machine collaboration · Collaborative robotics

1 Introduction

Nearly 40 years have passed since the presentation of the KUKA (http://www.kuka-robotics.com) FAMULUS, the first industrial robot to use the now standard setup of six electric motor-driven axes. In this timespan, industrial robots have become faster, more efficient, and more accurate—but most importantly also more affordable and accessible. Where the use of automation and robotic arms used to be the domain of large industries, now even small firms and offices can afford robotic arms and benefit from automation. This is not only made possible by the

S. Shepherd (✉)
KUKA Robotics Corporation, Shelby Township, MI, USA
e-mail: StuartShepherd@KUKARobotics.com

A. Buchstab
KUKA Roboter GmbH, Augsburg, Germany
e-mail: aloisbuchstab@kuka-roboter.de

W. McGee and M. Ponce de Leon (eds.), *Robotic Fabrication in Architecture, Art and Design 2014*, DOI: 10.1007/978-3-319-04663-1_26,
© Springer International Publishing Switzerland 2014

affordable prices, but also through new interfaces, sensors and strategies that safely enable the accessible programming of industrial robot, by e.g. allowing the direct import of G-code or utilizing new, visual programming strategies for defining robotic tool paths.

This development brings along new possibilities, but also new challenges. Where an automotive welding line is largely unattended, new applications may see robots in much closer proximity to, and collaboration with, human workers. Similarly, there is a great potential in allowing robots to work on-site or even to move themselves autonomously on-site, significantly increasing the flexibility of these multifunctional machines.

New technical solutions that go beyond current industrial robots will be required to realize these goals.

2 Lightweight Robots

The KUKA lightweight robot (LWR, Fig. 1) marks a departure from the way robots commonly work, towards a *production assistant* that combines sensor-based control with an excellent mass-payload ratio of 1:1, facilitating its mobility. This represents a significant step up from other industrial robots, which commonly achieve ratios of around 1:10.

It was initially developed at DLR for use in non-structured environments and interaction with humans (Albu-Schaeffer et al. 2007). Where common robotic arms depend on position control via position sensors on the motor side and current measurements, the KUKA LWR implements compliance control with position sensors on the motor side as well as position and torque sensors on the output side. Instead of relying on a stiff design of the mechanical structure, the LWR achieves high stiffness through active vibration damping.

In compliant mode, these sensors allow the user to program the robot manually—not through the KUKA control panel, but by physically moving the robot's joints (Bischoff et al. 2010). More importantly, the sensors enable the robot to continuously monitor itself and to stop when certain torque thresholds are exceeded or unexpected objects are sensed. In the case of a collision, where most other robots would seriously injure a person in its path, impact tests with the LWR show only a very low injury level even at its maximum speed of 2 m/s (Albu-Schaeffer et al. 2007). These results are also made possible by the special design of the LWR, relying on a soft silhouette with rounded edges to offer increased protection.

An additional benefit of the soft-robotics model is that it can be used to deal with inherent product- or material tolerances. With a regular, "stiff" robot, positioning data has to be completely accurate, requiring expensive equipment such as laser scanners. Inaccuracies can lead to damage of the material or even the robot, e.g. when a workpiece does not fit perfectly into a mounting element. Using

Fig. 1 KUKA LBR iiwa lightweight robot with seven axes

impedance control, the lightweight robot itself is able to react to resistance, e.g. by rotating or slightly shifting the element, instead of forcibly moving along the preprogrammed path. This system is successfully applied for the assembly of complex gears in the automotive industry.

Finally, its low weight and mass-payload ratio make the LWR a highly mobile robot—both in regard to portability, but also as a component of larger, robotic systems. The most well-known example being DLR's Justin, a humanoid robot that consists of a torso with three degrees of freedom, coupled with two KUKA lightweight robots as arms. Lightweight robots have also been mounted on mobile platforms such as the KUKA omniRob, especially in the field of service robotics. Finally, the low weight and effective vibration dampening make the LWR ideal for new applications, e.g. as robotic camera dollies, with firms such as CMOCOS (http://www.cmocos.com) already using lightweight-robot–mounted cameras for Hollywood movies (Fig. 2).

The most current iteration of the lightweight robot is the KUKA LBR iiwa—the intelligent industrial work assistant—that is available with payload capacities of 7 and 14 kg, making it the first lightweight robot for industrial use with a payload capacity of over 10 kg. It is also the first robot to implement KUKA.sunrise, a single controller kernel with modular, open interfaces, enabling object-oriented programming of complex robot systems, including safe, sensor-based, multi-kinematic systems.

Fig. 2 CMOCOS using a KUKA lightweight robot as a car-mounted camera dolly for the Dreamworks movie Need for Speed

3 On-Site Robotics

In the area of service robotics, the combination of robotic arms with a mobile platform to expand the robot's workspace (see Sect. 2) is very common. However, doing so with larger, industrial robots is complicated due to their high weight—the easiest way to increase a robotic arm's workspace is therefore the use of additional rotational or linear axes, which either rotate the workpiece or linearly move the robot itself. While these external axes are robust and proven technology, they again constrain the movement of the robot and cannot easily be moved or expanded.

The KUKA omniMove is a scalable, mobile platform for moving payloads of up to 90 t with an accuracy of ± 1 mm, based on an omnidirectional drive technology that allows it to move and rotate in any position. While it was initially developed for large-scale applications as in the aeronautical industry, the idea of mounting a large robotic arm on this platform was already shown in 2010. A similar movement concept is today already used for the compact KUKA youBot (Fig. 3).

In 2013, KUKA presented the moiros concept, which combined existing, state-of-the-art components to create an industrial robot with an unconstrained workspace, capable of working autonomously for over 8 h due to its high-powered, 20 kWh battery system.

Fig. 3 Up to 90 t payload KUKA omniMove (*left*), 0.5 kg payload youBot (*right*)

Fig. 4 KUKA moiros—mobile industrial robot system concept

Moiros (Fig. 4) consists of an 8 t payload omniMove platform, onto which the 32 individual batteries, a KR C4 generation controller, and a 120 kg payload robot of the KR QUANTEC series are mounted, offering 5 m of vertical and a nearly unlimited horizontal workspace, in which the robot navigates with the help of constant laser-scanning data.

A mobile system is therefore able to adapt itself to the size of the building elements and can be utilized flexibly, e.g. in different fabrication sites, depending on demand. Especially interesting is that the moiros concept only uses existing component that are already available on the market, making large-scale mobile robotics immediately possible.

In some disciplines such as architecture, there is significant interest in actually using robotic arms on construction sites—dusty, non-structured environments that

Fig. 5 Waterproof KUKA KR AGILUS WP

are exposed to the elements. The KUKA KR AGILUS Waterproof marks a significant step towards enabling such applications as it is not only protected against contact and ingress of dust, but also waterproof, due to additional seals in the exterior as well as stainless-steel covers that replace parts previously made of plastic (Fig. 5).

4 Outlook

In the past years, many new technologies were developed that will over time become industry standards. Especially the new KUKA LBR iiwa pioneers innovative collaborative robot concepts that are expected to enable projects that go far beyond of what we are using robots for at the moment.

With the ongoing demographic change, service robotics will not only find their way into our homes to support elderly care, but will also become increasingly important for small and medium enterprises, either as a third-hand for workers or as fully automated machines that perform separate tasks—but integrated into a shared space, not within an enclosed cell. However, the challenges for achieving this man–machine collaboration with robots are not only of a technical nature, but also regulatory issues, as many safety regulations still prohibit workers from entering a robot's workspace. Over time, new regulations will have to be found that take advanced safety measures into account.

Already, technology pioneered by the lightweight robot family of robot is beginning to filter down to the more common robot series such as the KUKA KR AGILUS, switching either an axis or a Cartesian direction to a soft mode.

Especially in the creative industry, where factories are not built around robots, but robots are now progressively introduced, factors such as user-safety, lightweight, and waterproofing will greatly facilitate many processes.

References

Albu-Schäffer A, Haddadin S, Ott C, Stemmer A, Wimböck C, Hirzinger G (2007) The DLR lightweight robot: design and control concepts for robots in human environments. Ind Rob: Int J 34(5):376–385

Bischoff R, Kurth J, Schreiber G, Koeppe R, Albu-Schäffer A, Beyer A, Eiberger O, Haddadin S, Stemmer A, Grunwald G, Hirzinger G (2010) The KUKA-DLR lightweight robot arm—a new reference platform for robotics research and manufacturing. In: Proceeding of: ISR/ROBOTIK 2010

ABB Robotic Technology in Art and Industry

Martin Kohlmaier, Nicolas De Keijser and John Bubnikovich

Abstract New robotic applications are enabled through the interlocking of developments in both robotic hardware and software. Specialized software interfaces and addons are turning multifunctional robotic arms into highly specialized and optimized machines for a variety of applications. In parallel robots are continuously improved through new materials, gears, and motors, leading to a higher reliability, accuracy, and speed, all at a lower energy consumption.

Keywords ABB · Robotic interfaces · Robot art · Green technology

1 Introduction

Industrial robots are considered to be one of the most important industrial technologies of our time. However, today's robotic users are no longer limited to large, industrial companies, but also include various small and medium enterprises, as well as even architects, artists and designers.

In the 40 years since ABB (http://www.abb.com) presented the first IRB6 robot in 1974, robotic arms have become faster, cheaper, more accurate and even more multifunctional, with the initial welding applications now being just one out of many use-cases for robotic arms. These advances are the result of research and development in both robotic hardware and software.

M. Kohlmaier (✉)
ABB AG, Robotics, Wr. Neudorf, Austria
e-mail: martin.kohlmaier@at.abb.com

N. De Keijser · J. Bubnikovich
ABB Robotics, Auburn Hills, MI, USA
e-mail: nicolas.de-keijser@us.abb.com

J. Bubnikovich
e-mail: john.bubnikovich@us.abb.com

W. McGee and M. Ponce de Leon (eds.), *Robotic Fabrication in Architecture, Art and Design 2014*, DOI: 10.1007/978-3-319-04663-1_27,
© Springer International Publishing Switzerland 2014

2 Advances in Robotic Software

As multifunctional machines, robotic arms can be applied in countless ways, from pick-and-placing to subtractive manufacturing and product testing. The main limitation is therefore not so much the hardware, rather than the software that enables an accessible programming of complex robotic tasks. In 1998, ABB published RobotStudio as the first simulation tool to use a virtual controller. Since then, the solution has been continuously expanded with a wide range of so called PowerPacs that add additional functionality to RobotStudio. PowerPacs include *ArcWelding* for welding applications, *Painting* for both the simulation of painting robots and prediction of paint process performance parameters (Fig. 1), *Bending* and *MachineTending* for cooperating with external machines, *Cutting* for optimized lasercutting, and *Palletizing*.

While these applications work offline, i.e. at an external computer, a range of add-ons is also available for the robot's FlexPendant HMI (human machine interface). Aiming towards smartphone simplicity, the user interface can be customized with specialized widgets and apps that are downloaded from RobotApps, offering special functionality for the user. For example, RobotWare Arc and Cutting complement the RobotStudio PowerPacs, while RobotWare Dispense controls processes for gluing, sealing, and extruding. Other add-ons work in the background, such as SoftMove, which represents a soft servo option that allows the robot to be compliant to external forces.

At the border between software and hardware are innovative products such as ABB's Integrated Vision and Integrated Force Control. Both consist of a sensor that can be directly integrated into the ABB environment. In the case of ForceControl, the sensor is placed between the robot and the endeffector and enables tactile sensing for the automation of complex tasks. Integrated Vision consists of an ABB Smart Camera that directly interfaces with the robot, enabling the robot to scan barcodes, measure surface defects, and to verify tolerances. Both robot and camera are programmed within the same RobotStudio environment, making interaction quick and intuitive.

3 Advances in Robotic Applications

The range of developments in robotic hardware and software have enabled completely new applications that go far beyond what was previously possible. Robotic arms are now capable of high-pressure cleaning massive mining trucks, while avoiding sensitive areas based on 3D-scanning data (Fig. 2). In the press automation sector, the TRX Twin Robot Xbar uses two cooperating robots to load and unload elements that would be too large for a single robot to handle.

Fig. 1 Spraypainting using ABB robots at Audi

Beyond industrial applications, new workflows and techniques are also developed for artistic projects, such as the "Long Distance Art" installation by artist Alex Kiessling (Fig. 3). Three ABB IRB 4600 robots were placed hundreds of kilometers apart at Vienna's Museumsquartier, London's Trafalgar Square, and Berlin's Breitscheidplatz. While the artist drew a large-scale portrait, an infrared touch-frame coupled with a Kinect sensor captured the movement of the pen and translated it into robot commands. These commands were then streamed to all three locations, with the robots creating their interpretation of the drawing in real time.

4 Green Technology

Not all advances in robotic technology are as visible as in the projects in Sect. 3, but offer enhancements that may be even more significant.

The new IRB 6700 marks the 7th generation of large ABB robots and features a multitude of next-generation improvements. Rated for 400,000 h (MTBF) it is more robust than its predecessor and maintenance has been simplified, making it the highest performing robot for the lowest total cost of ownership in the 150–300 kg class. Part of this efficiency is also reflected in its green and sustainable design, using 15 % less energy than an IRB 6640.

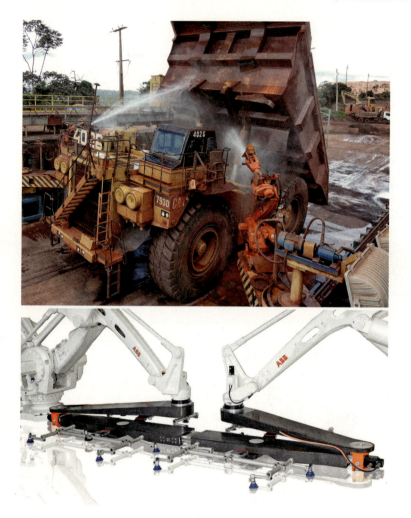

Fig. 2 Robotic truck washing (*above*), ABB TRX with cooperating robots for loading and unloading of large elements (*below*)

It also solves the frequent robot-problem of tangling process cables. Lean ID—the new Integrated Dressing (ID) solution—routes cables inside the upper arm and along the robot wrist. Thus, cables follow every motion of the robot arm, instead of coming into swing in irregular patterns—making the robot simulation more accurate and prolonging its life time.

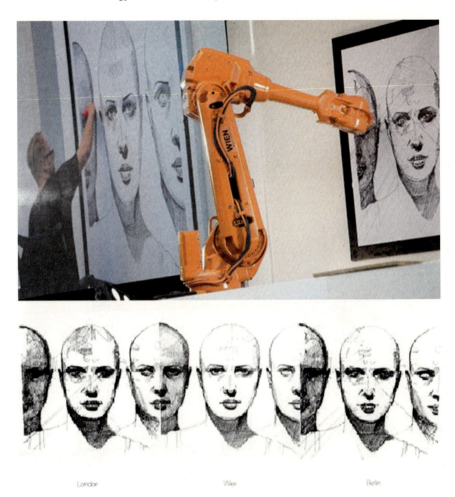

Fig. 3 "Long Distance Art" by Alex Kissling (http://www.alexkiessling.com) with INDAT (http://www.indat.at) and Strukt [4] http://www.strukt.com

5 Outlook

Robotic technology is constantly advancing, both in highly visible steps, but also in steps whose impact may only be felt in the future. Through the development of new tools and new software at ABB, automation is being made available to an increasing number of industries, from large-scale enterprises to a single artist.

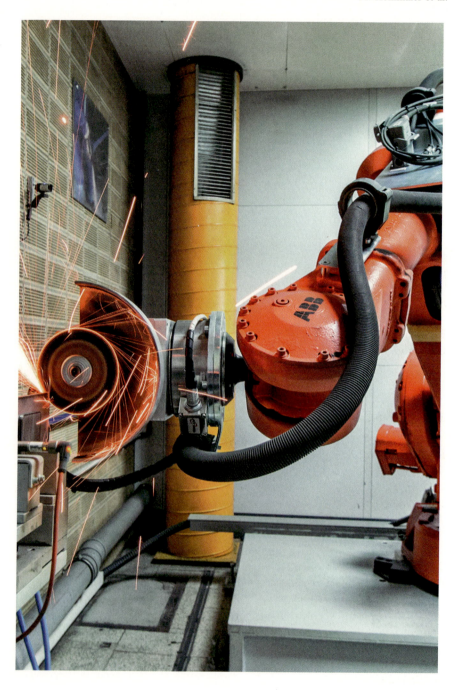

Special Solutions for Special Applications

Joe Gemma and Manfred Hubschmann

Abstract The use of robotic arms is starting to move far beyond its roots at the automotive industry, with robots now being used in increasing numbers in the general industry, as well as the creative industry. New applications continue to pose new challenges, which can no longer be fulfilled with standard robotic arms. Specialized robot types by Stäubli now offer properties such as water resistance and integrate milling or spray painting equipment directly inside their arms.

Keywords Stäubli · Specialized robots · High speed milling · Spray painting · Clean room

1 Introduction

Robotic arms are highly complex and sophisticated machines, whose main strength lies in their multi-functionality. Thus, most robots are not built specifically for a clearly defined application, but rather as flexible device whose mechanic purpose can be changed simply by mounting a different tool. However, not all robots are set up in areas with standard temperatures and standard humidity or perform tasks that could be performed by any type of robot.

Alongside the standard range of six-axis robotic arms, ranging from 1 to 250 kg, Stäubli (http://www.staubli.com) has been developing specialized series of robots to fulfill very specific requirements, bringing robotic arms into completely new areas of industry and research.

J. Gemma (✉)
Stäubli, Duncan, USA
e-mail: j.gemma@staubli.com

M. Hubschmann
Stäubli, Bayreuth, Germany
e-mail: m.hubschmann@staubli.com

W. McGee and M. Ponce de Leon (eds.), *Robotic Fabrication in Architecture, Art and Design 2014*, DOI: 10.1007/978-3-319-04663-1_28,
© Springer International Publishing Switzerland 2014

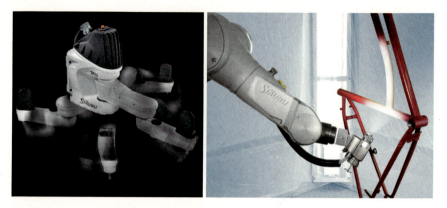

Fig. 1 Stäubli TP80 fast picker (*left*), Staeubli painting robot (*right*)

2 High-Speed Without 6 Axes

In the creative industry, most robotic arms use a six axis kinematic layout. While this allows a maximum of flexibility as well as a spherical work envelope, specialized applications such as high-speed pick and placing require extremely high speeds and low cycle times within a vertically limited workspace. The Stäubli TP80 (Fig. 1, left) fast picker is specifically tailored towards such applications, moving its four axes with the highest possible speed through its 800 mm reach, while still achieving ±0.05 mm repeatability. Within just a minute, 200 elements can be transferred by more than 700 mm each.

3 Painting Robots

Robotic spray painting offers significant benefits over manual labor, as the robot's high accuracy and constant speed lead to a very regular and optimized finish with a minimum of wasted paint (Fig. 1, right). While spray painting does not require a particularly high payload, a large work envelope and resistance to solvents are crucial. The TX250 paint robot combines a reach of more than 2.5 m with a very compact profile and just 250 kg of weight. Additionally, it offers internal valves and pumps as well as a hollow wrist so that all painting equipment can fit *inside* the arm, reducing the risk of contamination. To facilitate the programming of painting and coating processes, the TX250 uses the PaintiXen software which provides simplified programming interface. These concepts can not only be applied to painting, but also to enameling, coating, glue spreading, and glazing.

Fig. 2 Stäubli RX170 HSM
robot for high speed
machining with integrated
milling spindle

4 Robotic Milling

Milling has become an important application for industrial robots, offering a larger workspace at a lower cost than regular multi-axis CNC machines. The Stäubli RX170 HSM (Fig. 2) is the first dedicated high-speed machining robot, made with a special attention to the machining of oversized CFRP parts. By integrating the spindle into the forearm of the robot, the RX170 HSM achieves improved rigidity and precision as well as reduced slanting. Three different spindle models are available, covering a range of 8–17 kW and 500–42,000 rpm. Similar to the TX250 paint robot, all cables run inside the arm, and a special software (VALhsm) is available to facilitate typical milling or drilling processes.

5 Clean and Waterproof

Special environments also pose new challenges for robotic arms, such as humid or explosive atmospheres, machine-tool environments, and cleanrooms. Stäubli's HE range of robots (Fig. 3) are optimized for humid environments such as water jet cutting, cleaning, and food processing. They use an enclosed, waterproof arm

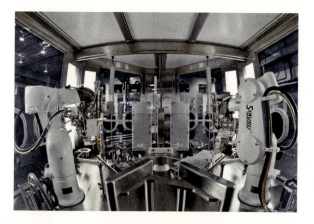

Fig. 3 Stäubli HE robots in a cleanroom environment (*left*), protected interconnection cables for robots in humid or cleanroom environments (*right*)

structure and protect the interconnection cables by placing them underneath the arm itself. A smooth and shiny finish makes them easy to clean on the outside, while the optional pressurization of the arm helps keeping liquids as well as explosive vapors, chips, and oil outside. Even better sealing is achieved by the stericlean range of robots, which are optimized for a minimum of emissions in cleanrooms and made to withstand being cleaned with gaseous hydrogen peroxide.

6 Outlook

Specialized robotic solutions represent a substantial step towards making automation available to new industries. In addition to progress in the field of robotic hardware, software developments such as Stäubli VALhsm and PaintiXen are similarly required to make robotic arms actually accessible to new users.

New fields of robotic users such as the creative industry show that the full potential of robotic arms is still untapped.

Sensitive Robotic Processes

Advances in Gripper Technology

Christian Binder

Abstract Robotic arms are highly multifunctional machines. A large part of the versatility is due to gripping technology, as grippers enable the robot not only to pick-and-place objects, but also to handle a wide variety of tools, and to manipulate material, e.g. via polishing and grinding. New developments by Schunk further increase the potential of grippers, via customized, 3D-printed gripper fingers, highly sensitive force-moment sensors, and completely new gripping tools, equipped with tactile sensors and their own intelligence.

Keywords Gripping technology · Rapid prototyping · Force-moment sensors · Schunk

1 Introduction

Customization is becoming increasingly important in the field of robotic fabrication, from the automotive industry to new fields such as architecture, art, and design. Where the core competence of robotic arms used to be the untiring, accurate repetition of tasks, nowadays tasks are often much more elaborate and demand a high degree of flexibility.

The benchmark of flexibility in gripping technology is of course the human hand, having an excellent weight/payload ratio and both a fine touch and high strength. Replicating this feat has been the goal of engineers for a long time, and new technologies developed by Schunk (http://www.schunk.com) are now starting to make it happen.

C. Binder (✉)
SCHUNK Intec GmbH, Traun, Austria
e-mail: Christian.Binder@at.schunk.com

W. McGee and M. Ponce de Leon (eds.), *Robotic Fabrication in Architecture, Art and Design 2014*, DOI: 10.1007/978-3-319-04663-1_29,
© Springer International Publishing Switzerland 2014

Fig. 1 Customized, light and wear-free Schunk gripper fingers made of polyamide

2 Flexible Gripper Design

A regular, industry standard gripper consists of an actuator and two gripper fingers with a defined stroke. Such setups have been used for decades and provide a very affordable and proven way of quickly pick-and-placing elements. However, these grippers work best with simple geometries that are easy to grasp, e.g. with parallel edges. Problems arise with more complex shapes that cannot be grasped with two parallel fingers. A possible solution may be more complex gripping setups, such as the use of three angular fingers, or gripper fingers that are custom-made for a particular object geometry—involving a tooling process that used to be complex and expensive.

However, the rise of 3D-printing and additive manufacturing now enables Schunk to make the full customization of gripper fingers possible, even for prototypical projects and small production runs (Fig. 1). Designers can create gripper fingers with an enormous degree of freedom, without having to pay attention to draft angles or undercuts—it is even feasible to integrate channels for power, signal, or compressed air feeding directly in the gripper, and to manufacture moveable parts such as hinges in a single piece, without elaborate assembly. The laser-sintered, light, and wear-free polyamide offers cost-savings in regards to its manufacturing, as well as in regards to the day-to-day use of the gripper, reducing weight and thereby cycle time.

3 Feeling and Gripping

However for some tasks, it is not enough to just safely grasp an object. Especially for processes that are not fully predictable, the gripper has to actually feel the forces that are applied to an object, e.g., for machining tasks such as grinding and

Fig. 2 Force moment sensors (*left*), mounted between the robot's flange and gripping tool for grinding (*right*)

polishing. As sensors within the robotic arms are commonly not sensitive enough to recognize fine material changes, Schunk developed specialized force-moment sensors that are placed between the actuator and the robot's flange (Fig. 2). These sensors measure forces and moments in all six degrees of freedom up to 7,000 times a second, enabling the safe handling of fragile parts, or compensating for tool wear and inaccuracies in workpiece positioning.

4 SDH-2 Gripping Hand

Merging a flexible kinematic layout that allows it to grasp an extremely wide range of objects with a multitude of highly accurate sensors, the Schunk SDH-2 is the first industrial gripping hand with real fingertip feel. Its three double-joint fingers can be configured to perform among other things, the "three-finger centric", "two-finger parallel" and "cylindrical grip" industrial gripping operations as well as numerous other variations (Fig. 3).

Six tactile sensor fields register the contact forces on the gripping surfaces in a space-saving manner. They allow the hand to identify completely different objects and also handle similar parts in a secure and sensitive manner. As a result, the hand is able to grip reactively, since sensors identify whether an object is being held optimally or whether the grip has to be corrected. Furthermore, it is able to position completely different objects in order to join them. The intelligence of the

Fig. 3 SDH2 gripping operations

gripping module lies in the "wrist": The control strategy required for the particular gripping scenarios can be loaded into the memory of the hand's electronic control unit as a decentralized program module. The gripping hand also has a number of mechanical special features. For example, the connecting points and joints are statically and dynamically sealed and, in this way, protected against dust and moisture. To ensure a high degree of passive safety, the hand has no corners or sharp edges. Special rotary feed-throughs within the sealed fingers protect the entire cabling. The gripping speed and force can be programmed for specific tasks and processes so that gripping does not pose any danger. If a finger encounters an obstacle nevertheless, the drives in the hand detect the entailed increase in power consumption within a matter of milliseconds and the hand reacts accordingly.

5 Outlook

The customization of fabrication processes requires not only new programming strategies, but also new and flexible hardware that can be adapted for a wide range of applications. From standard parallel grippers that can be customized via 3D printing, to highly-complex robotic hands that can feel what they are grasping, Schunk offers a huge range of options to cover even the most special demands.

The Power of Engineering, the Invention of Artists

Bot & Dolly

Kendra Byrne, Jonathan Proto, Brandon Kruysman and Matthew Bitterman

Abstract Bot & Dolly is a design and engineering studio that specializes in automation, robotics, and filmmaking. It is our mission to advance technology as a creative medium, and build world-class tools that put robotics directly in the hands of creators. Our software and hardware platform enables simple, natural, and intuitive control over robotic automation as an integral tool in the creative and development process from rapid prototyping to production. Creating a toolset that allows motion for industrial robots to be designed with a high degree of temporal resolution, not just spatial resolution, broadens the potential for creative applications in robotics. To date our software platform has driven innovative applications of robotics across the media industry, including feature films, national television ads, Las Vegas shows, and large-scale art installations.

Keywords Synchronous robotics · Visual effects · Interactive automation

1 Introduction

When Bot & Dolly entered the industrial robotics space, the first challenge was clear—how does an artist, a builder, or any non-roboticist begin to engage with a toolset developed for a small group of specialists. The focus of our development

K. Byrne (✉) · J. Proto · B. Kruysman · M. Bitterman
Bot & Dolly, San Francisco, CA, USA
e-mail: kendra@botndolly.com

J. Proto
e-mail: jonathan.proto@botndolly.com

B. Kruysman
e-mail: brandon.kruysman@botndolly.com

M. Bitterman
e-mail: matthew.bitterman@botndolly.com

W. McGee and M. Ponce de Leon (eds.), *Robotic Fabrication in Architecture, Art and Design 2014*, DOI: 10.1007/978-3-319-04663-1_30, © Springer International Publishing Switzerland 2014

399

efforts has been advancing human robot interfaces to transform the notion of a roboticist and open automation technology up to vastly different creative spaces. By abstracting robot programming tools into a natural language, we have constructed a lens to look at robotics subjectively as creatives. We are adding to the traditional keywords that define the state of robotics—feedback loops, torque based models of control—with words that describe special effects and human experience. From an engineering perspective, this means the measure of resolution is extended from arc-seconds to frames-per-second. For designers, makers, and creatives this means opening up new ways to engage with robotics and new forms of artistic expression.

2 Technology and Impossible Effects: "Gravity"

"It can't be done." This is what Alfonso Cuarón heard when speaking of his vision of weightlessness for the film Gravity (Fig. 1). While these words discourage some, they inspire others. As a multidisciplinary team of designers and engineers, Bot & Dolly took these words as a challenge to choreograph interactions between the physical world and computational world that can only be described through time.

Bot & Dolly's work on Gravity began mid-2011 after the Head of Visual Effects at Warner Brothers reached out to the studio. He and his team were looking for the technology needed to bring Cuarón's film to life. Cuarón and Webber took a unique approach to simulating weightlessness. Instead of moving an actor through a set using standard methods of rigging, the illusion of zero-gravity was achieved by moving the world around the actor (Fig. 2). This could only be done by synchronizing lighting, backgrounds, and actor pose with frame accurate camera positioning, which is outside the capabilities of standard film production workflows and industrial automation toolsets.

Utilizing time as the primary driver of motion enables precise synchronization between robots, lights, and backgrounds by unifying all of the technology on set within a common timeline. Designed to control camera movement, activate lights, and shift set pieces with ever-repeatable precision, our robotics platform allowed the virtual and physical worlds to unite and CG elements to match in real time. Being able to execute motion that is frame accurate to the animation environment means that the physical world can be synchronized with the digital world. By leveraging the visualization capabilities of Autodesk Maya with a robotics platform precise to both the millimeter and the millisecond we were able to make an impossible vision a reality.

Bot & Dolly provided Cuarón and Framestore, an international visual effects company, with the necessary time based tools to execute complex cinematography based on computer previsualizations, in live action with industrial robots and other onset hardware. Because our tools integrate tightly with industry standard software Maya, Framestore was given control of camera, lighting and other set elements from within their established animation environment. The entire story of Gravity,

Fig. 1 "Gravity", Warner Brothers Production, 2013

Fig. 2 "Gravity", behind the scenes, 2010

down to the detail of lights and lenses, was previsualized prior to filming. Using Bot & Dolly's system of authoring and control, zero-gravity effects created in animation could be reproduced in the physical world.

Gravity was the culmination of leveraging visual effects designers and robotics engineers to make the physically impossible visually possible—to invert gravity on earth.

Fig. 3 "Remember reach", 2011

3 Technology and Participatory Experience: "Reach"/ "Kinetisphere"

The potential for robotic technologies in networked, participatory experiences engages a new level of interaction between the user and physical objects. In the following examples, a combination of both software and hardware solutions are exploited for new real-time interactive experiences that merge the virtual and the physical.

'Remember Reach' is a robotically driven, interactive tribute to the fallen members of Noble Team, characters from the HALO video game franchise (Fig. 3). Based on the notion of remembrance, this long exposure light sculpture ran continuously for 20 days placing 118,422 points of light whenever a new user visited the official Halo Reach website.

Reach represents possibilities of cloud based control of industrial robotic arms, enabling multi user participation from all over the world. All though in this instance the cloud based model is characterized by its passive interactions between users and the installation; this method of control allows us to rethink traditional linear methods of robot programming in favor of more dynamic and adaptable workflows.

Created to celebrate the launch of the Nexus Q, Kinetisphere (Fig. 4) is a monument to the confluence of modern design and digital music. Like the Nexus Q, Kinetisphere is designed to encourage social interaction and connect people. It's a multi-sensory, multi-user participatory experience. Motion, visual effects, and sound are created through gaming with the Nexus Q device, which acts as a flexible user interface to move the physical world. Instead of designing specific

Fig. 4 "Kinetisphere", Google I/O conference 2012

motion and effects sequences, we designed a baseline and defined allowable offsets and amplifications that a user could modulate using the Nexus Q device.

Kinetisphere debuted at the 2012 Google I/O conference, giving attendees the opportunity to control an 8-foot, 300-pound Nexus Q replica attached to the end of an industrial robot arm through three stations, each consisting of a Nexus Q device and a Nexus 7 tablet. One station controlled the height of the sphere, another its rotational angle about the end effector and the third controlled its rotational angle about its base plane. As the Kinetisphere was guided through its 3-dimensional work envelope users experienced both visual and audible amplification, culminating at virtual hotpots. These virtual zones placed throughout the installation encouraged users, through exploration and discovery, to craft their own unique experiences enabled by tangible devices.

4 Technology and Magic: "Box"

> Any sufficiently advanced technology is indistinguishable from magic.
> Arthur C. Clarke

Magicians combine the art of performance with scientific truths, including optics and physics, to create impossible illusions. Box is a short film, written and produced by Bot & Dolly, documenting a live performance grounded in the principles of Stage Magic: Transformation, Levitation, Intersection, Teleportation, and Escape. Through each act, Box explores the synthesis of real and digital space through projection-mapping onto moving surfaces controlling optics in both space and time (Fig. 5).

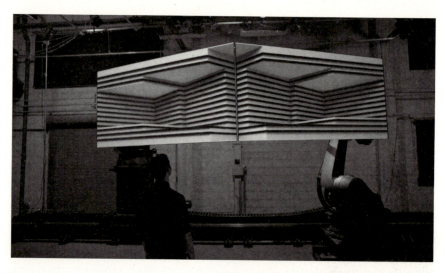

Fig. 5 "Box", projection mapping on moving surfaces, 2013

Box is the culmination of multiple technologies, including frame accurate industrial robotics, synchronized projection mapping, and software tools designed for rapid iteration of motion design. Motion art was designed and projected into a physical installation, where the illusions were captured live without any post production. The precision and fidelity of the technology masked the methods used to create the performance from the viewer making the technology indistinguishable from magic. Space itself could appear, fold and disappear. Flat objects could become portals for escape. Every component of each technology in the production was synchronized in both time and space to produce immersive optical illusions.

As a live experience, Box creates a physical world that people can actually walk into and see. We believe this methodology of collapsing the gap between the physical and the digital has tremendous potential to radically transform theatrical presentations, and define new genres of expression.

5 The Power of Engineering, the Invention of Artists

As engineers, one understands the importance of precision as it relates to control loops and calibration; as designers, precision is leveraged as a generative tool. Through the combination of these methodologies it means we know not only what tools are useful and what workflows are intuitive, but we can also imagine what robotics makes possible. We can make the physically impossible visually possible by controlling time and light as visual designers. As architects, we can blend space between the unpredictabilities of a physical site and the exactness of a digital model.

Robots in Architecture 2014 Scientific Committee

Sean Alhquist, Taubman College of Architecture and Urban Planning, University of Michigan, USA

Kristy Balliet, Austin E. Knowlton School of Architecture, Ohio State University, USA

Tobias Bonswetch, Rob Technologies AB, Switzerland

Johannes Braumann, Robots in Architecture, Austria

Sigrid Brell-Cokcan, Robots in Architecture, Austria

Jan Brueninghaus, Faculty of Mechanical Engineering, TU Dortmund, Germany

Brandon Clifford, MIT Architecture, USA

Jason K. Johnson, California College of the Arts, USA

Axel Kilian, Princeton University School of Architecture, USA

Branko Kolarevic, Faculty of Environmental Design, University of Calgary, Canada

Andrew Payne, LIFT architects, USA

Gregor Steinhagen, Faculty of Mechanical Engineering, TU Dortmund, Germany

Larry Sass, MIT Architecture, USA

Martin Trautz, Institute of Supporting Structures, RTWH Aachen, Germany

Glenn Wilcox, Taubman College of Architecture and Urban Planning, University of Michigan, USA

Aaron Willette, Taubman College of Architecture and Urban Planning, University of Michigan, USA

W. McGee and M. Ponce de Leon (eds.), *Robotic Fabrication in Architecture, Art and Design 2014*, DOI: 10.1007/978-3-319-04663-1,
© Springer International Publishing Switzerland 2014